ATMOSPHERIC BIOGENIC HYDROCARBONS

Volume 2

Ambient Concentrations and Atmospheric Chemistry

ATMOSPHERIC BIOGENIC HYDROCARBONS

Volume 2

Ambient Concentrations and Atmospheric Chemistry

Edited by
Joseph J. Bufalini
Robert R. Arnts

ANN ARBOR SCIENCE
PUBLISHERS INC / THE BUTTERWORTH GROUP

Library of Congress Catalog Card Number 81-52298
ISBN 0-250-40498-2

Manufactured in the United States of America

Butterworths, Ltd., Borough Green, Sevenoaks, Kent TN15 8PH, England

PREFACE

These two volumes discuss natural organic emissions and their impact on the atmosphere. Although often termed "natural pollution," these emissions are naturally occurring, and thus cannot be considered pollutants at all.

Estimates of the types and amounts of volatile organic compounds emitted by plants to the atmosphere were made in the early 1960s by F. W. Went. Interest in these naturally emitted organics increased dramatically in the late 1960s, when scientists at the Stanford Research Institute estimated that global natural organic emissions (excluding methane) far exceeded anthropogenic emissions. Since that time, research has accelerated to refine emission estimates, to measure ambient air hydrocarbon concentrations, and to study the atmospheric chemistry of these compounds. The results of such investigations are of interest to regulatory agencies concerned with formulating scientifically sound air pollution control strategies as well as to atmospheric chemists studying the composition of "clean" atmospheres.

The two volumes are the result of a January 8–9, 1980, symposium on Atmospheric Biogenic Hydrocarbons sponsored by the U.S. Environmental Protection Agency (EPA), which was held at EPA's Research Center, Research Triangle Park, North Carolina. The symposium brought together scientists to address specific aspects of natural organic emissions. Invited speakers represented the fields of botany, ecology, plant physiology, micrometeorology, forestry and atmospheric chemistry; each has actively been researching the issues presented herein. As such, these volumes represent the first attempt to gather the disciplines which must be applied to assess the accuracy of the emission estimates and to estimate the impact of biogenic emissions on the chemical composition of the atmosphere. The reader should find the discussions following the formal papers particularly enlightening. These discussion sections were based on verbatim transcripts made at the conference. Any comments or information inserted in these sections by the editor are enclosed within brackets.

Joseph J. Bufalini
Robert R. Arnts

ACKNOWLEDGMENTS

We are indebted to the authors and participants in the symposium for their invaluable time and efforts. Also, we commend Northrop Services, Inc., of Research Triangle Park, for their fine work in the planning and execution of the symposium. In particular, we are indebted to Ms. Jill D. Snider of Northrop for coordination of the symposium and for her patience and editorial skills in preparing this book for publication.

Joseph J. Bufalini is chief of the Gas Kinetics and Photochemistry Branch at the EPA Environmental Sciences Research Laboratory, Research Triangle Park, North Carolina. Since 1961 he has been involved in various aspects of air pollution research, including the study of mechanisms and rates of reactions of gaseous pollutants, study of intermediate and final oxidation products and modeling studies for hydrocarbon-nitrogen oxide systems.

Dr. Bufalini received his BS at Siena College, Albany, New York, and his MS and PhD in physical chemistry from the University of Arkansas, Fayetteville, Arkansas. He has authored many articles and papers in scientific journals, and has received the bronze medal for government superior service.

Robert R. Arnts is a chemist in the Gas Kinetics and Photochemistry Branch at the EPA Environmental Sciences Research Laboratory, Research Triangle Park, North Carolina. Since 1971 he has been active in various field and laboratory studies concerning atmospheric chemistry. These include studies of the chemistry and ambient measurement of halogenated organics, characterization of the atmospheric chemistry of isoprene and the monoterpenes, and application of micrometeorological techniques to estimate biogenic hydrocarbon emissions.

Mr. Arnts holds a BS in Chemistry from Moravian College and an MS in Environmental Science from Rutgers University.

CONTENTS

Volume 1

Volume 2

LIST OF ABBREVIATIONS, SYMBOLS AND TRADEMARKS

ABBREVIATIONS

AQCR	Air Quality Control Region
ARL	Atmospheric Research Laboratory
CDS	Compliance Data System
CNC	condensation nuclei counter
CRC	Coordinating Research Council
DBH	diameter at breast height
DVM	digital volt meter
EAA	electrical aerosol analyzer
EHIS	Emissions History Information System
EKMA	Empirical Kinetic Modeling Approach
EPA	Environmental Protection Agency
ERT	Environmental Research and Technology, Inc.
ESRL	Environmental Sciences Research Laboratory
F-11	Freon-11
F-12	Freon-12
FEP	fluorinated ethylene propylene
FID	flame ionization detector
FS	fluorescence spectrophotometry
FT-IR	Fourier transform-infrared
GC	gas chromatography
GM	General Motors
HAOS	Houston Area Oxidant Study
HATREMS	Hazardous and Trace Emissions System
HC	hydrocarbon
HCE	hydrocarbon emissions
HID	high-intensity discharge
IBP	International Biological Program
IC	ion chromatography
IR	infrared
JSF	Jones State Forest
LAI	leaf area index
MBTH	3-methyl-3-benzothiazolone hydrazone
MCA	multichannel analyzer
MS	mass spectrometry
NCAR	National Center for Atmospheric Research

NEDS	National Emissions Data System
NMHC	nonmethane hydrocarbon
NMR	nuclear magnetic resonance
OAQPS	Office of Air Quality Planning and Standards
OH	hydroxyl radical
OPC	optical particle counter
PAH	polycyclic aromatic hydrocarbon
PAR	photosynthetically active radiation
PGA	phosphoglyceric acid
PP	palisade parenchyma
RTI	Research Triangle Institute
RTP	Research Triangle Park
SIP	State Implementation Plan
THC	total hydrocarbon
TIE	The Institute of Ecology
TLC	thin layer chromatography
TNHC	total natural hydrocarbon
TNMHC	total nonmethane hydrocarbon
TSP	trisodium phosphate
UNC	University of North Carolina
UV	ultraviolet
VOC	volatile organic compound
VMT	vehicle miles traveled
XRF	X-ray fluorescence

SYMBOLS

CCl_4	carbon tetrachloride
CH_3OH	methanol
CH_3CH_2OH	ethanol
$CH_3(CH_2)_2OH$	propanol
$CH_3(CH_2)_3OH$	butanol
$CH_3(CH_2)_4OH$	pentanol
CH_4	methane
C_2H_2	acetylene
CO	carbon monoxide
CO_2	carbon dioxide
H^+	hydrogen ion
H_2	hydrogen gas
HNO_3	nitric acid
H_3PO_3	phosphorous acid
H_2S	hydrogen sulfide
H_2SO_4	sulfuric acid
K_2CO_3	potassium carbonate
NH_4^+	ammonium sulfate
NO	nitric oxide

NO_2 nitrogen dioxide
NO_3^- nitrate ion
NO_x nitrogen oxides
O_2 oxygen gas
O_3 ozone
SF_6 sulfur hexafluoride
SO_2 sulfur dioxide
$SO_4^=$ sulfate ion
SO_x sulfur oxides

TRADEMARKS

Mitex   A registered trademark of the E. I. du Pont de Nemours & Co., Wilmington, Delaware

Mylar   A registered trademark of the E. I. du Pont de Nemours & Co., Wilmington, Delaware

Pyrex   A registered trademark of Corning Glass Works, Corning, New York

Teflon   A registered trademark of the E. I. du Pont de Nemours & Co., Wilmington, Delaware

Tenax   A registered trademark of Enka, N. V., The Netherlands

Tedlar   A registered trademark of the E. I. du Pont de Nemours & Co., Wilmington, Delaware

10. NON-URBAN HYDROCARBON CONCENTRATIONS
IN AMBIENT AIR NORTH OF HOUSTON, TEXAS

Robert L. Seila. Environmental Sciences Research
Laboratory, U.S. Environmental Protection Agency,
Research Triangle Park, North Carolina   27711

ABSTRACT

    In January 1978, a study was undertaken at W. G. Jones
State Forest, 38 miles north of Houston, Texas, to: (1)
determine the concentrations of nonmethane hydrocarbons,
methane, and carbon monoxide; (2) detail the composition of
hydrocarbons (especially of the vegetation); and (3) discover
the sources of nonmethane hydrocarbons. Thirteen, 3-h
integrated Tedlar bag samples and five grab samples using
stainless steel cans were collected over a 30-h period. The
samples were returned to the Research Triangle Park
laboratory for analysis, where the can samples showed lower
nonmethane hydrocarbon concentration values than did the bag
samples. Sources of paraffins (72% of the nonmethane
hydrocarbons) and the other hydrocarbons were found to be:
vehicular exhaust (35%), the forest's vegetation (2%), the
city of Houston (22%), and the region between Houston and the
forest (32%). Isoprene and α-pinene were the vegetative
hydrocarbons noted, with the latter showing a distinctive
24-h cycle of concentration variation.

INTRODUCTION

    The importance of vegetative hydrocarbons (HC's) in the
photochemical production of ozone (O_3) is a subject of
ongoing controversy and debate. The issue has been defined
and discussed by Dimitriades and Altshuller (1977, 1978), and

reviewed and analyzed by Coffey and Westberg (1977). The
data concerning photochemical reactivities, emission rates,
and ambient concentrations of vegetative HC's are the
evidence on which analyses and opinions are based. This
report concerns the ambient concentration component of the
issue.

Whitehead and Sievers (1977) reported a mean ambient
total nonmethane hydrocarbon (TNMHC) concentration of 8.7 ppm
for 35 morning samples collected at W. G. Jones State Forest
(JSF), a 1700 acre tract located in Montgomery County 38
miles north of Houston, Texas. They concluded that the "high
TNMHC levels observed were produced by the forest vegetation." Their results seem to run counter to the prevailing
evidence concerning ambient vegetative HC concentrations.
Westberg reviewed the available ambient concentration data
and concluded that natural HC concentrations are on the order
of a few ppb carbon, even in forested regions (Coffey and
Westberg 1977). The ambient α-pinene concentrations measured
by this laboratory at a loblolly pine plantation in central
North Carolina were only a few ppbC, ranging from 0.6 ppbC to
13 ppbC for over 300 samples collected above the canopy
during midday. The highest concentration observed within the
canopy itself was 55 ppbC during night inversion conditions.
Even when limbs were enclosed in Teflon bags--a method for
determining vegetative emission rates (Zimmerman 1979)--the
TNMHC concentrations rarely exceeded 8.7 ppm.

Since Whitehead and Sievers used a total HC analyzer for
their investigation, they had no means for determining the
specific HC's contributing to their TNMHC values. In order
to determine the specific HC's and their contributions to the
TNMHC burden at JSF, a 3-day sampling program was undertaken.
This report presents the results of that study.

EXPERIMENTAL SECTION

Sampling

Jones State Forest is located 1 mile west of Interstate
Highway 45 on Farm to Market Road (FM) 1488, and 5 miles
southwest of Conroe, the nearest town. It is operated by the
Texas Forest Service for research and recreation (picnicking
and camping). The predominant vegetation is loblolly pine.

The JSF sampling consisted of diurnal sampling using
bags and grab sampling using steel cans. Thirteen 3-h
integrated samples were collected in Tedlar bags from January

2

4 to January 6, 1978. Five grab samples were collected in stainless steel cans on January 5. In addition to the JSF sampling, two samples of commercial natural gas from the Houston Medical Center were collected in cans during March 1978.

Samples were collected at two sites within JSF. All of the thirteen diurnal samples and three of the can samples were collected at a picnic area about 100 m from FM 1488. This site was the only location with electricity for operating a continuous sampling system, yet it was not an ideal site because of the close proximity of FM 1488, a rural paved road with moderate traffic. However, southerly winds kept the road downwind of the picnic site during the entire study, which minimized the impact of local automobile pollution. The second site was deeper in the forest away from the potential impact of the road. Two can samples were collected there by means of a battery operated pump.

Diurnal sampling was performed from 1500 CST, January 4, to 0600 CST, January 6, 1978. A sequential sampler was used to collect 3-h integrated samples in 10 1 capacity 2 mil Tedlar bags; construction of the bags is described elsewhere (Seila et al. 1976). The bags were leak-tested by evacuation prior to shipment to Houston, and by subsequent visual inspection immediately before use. If air had leaked into a bag, it was not used. In order to prevent reaction with the HC's in the samples, each bag was spiked just prior to attachment to the sampler with 15 ml of a 38 ppm mixture of nitric oxide (NO) in nitrogen. This mixture destroyed any O_3 present in the air being sampled. Only 5 1 of sample were collected to allow for expansion during the return flight to Research Triangle Park. Upon return, three very deflated bags were presumed to have leaked and were not analyzed.

Five internally electropolished 4.5 1 stainless steel cans were used for the grab samples to provide a comparison with the bag samples. The cans were purged with approximately 50 1 of pre-purified nitrogen before shipment to Texas. Three can samples were collected by purging the cans with ambient air for 5 min and then pressurizing them to approximately 15 psi, using a Teflon diaphragm dc pump energized by a 12-V car battery.

Analysis Methods

The JSF samples were analyzed for methane (CH_4), carbon monoxide (CO), and C_2 to C_{10} individual HC's; the natural gas samples were analyzed for C_1 to C_5 HC's. Gas chromatography

3

(GC) was the analytical method used for analysis of the C_2 to C_{10} HC's. The system consisted of three columns on three flame ionization detectors with cryogenic preconcentration of the sample. Details of this system are described elsewhere (Seila 1979). Methane and carbon monoxide determinations were performed on a Beckman 6800 air quality chromatograph. The natural gas analyses were performed on the same silica gel column that was employed for the ambient C_2 to C_5 aliphatics analyses. However, cryogenic preconcentration was not required for the natural gas analyses. All GC detector outputs were recorded on strip chart recorders, while an electronic digitizer was used to measure peak areas and heights from the strip chart chromatograms. Response factors determined from analyses of known concentration compounds were used to convert the area or height measurements to concentration.

RESULTS AND DISCUSSION

HC/CO Results

A summary of the TNMHC, CH_4, and CO results is presented in Figure 1, and a detailed breakdown of the individual HC concentrations is available in a separate report (Seila 1979). The TNMHC values, derived by summing the individual HC concentrations, ranged from 124 ppbC to 531 ppbC. Although the mean CO concentration for this study, 0.63 ppm $\pm$ 0.16 ppm, compares favorably with Whitehead's mean, 0.60 ppm $\pm$ 0.15 ppm (Whitehead and Sievers 1977), the HC results are very disparate. The mean TNMHC for the bag samples was 0.306 ppm $\pm$ 0.096 ppm, and 0.184 ppm $\pm$ 0.069 ppm for the can samples, compared to Whitehead's mean of 8.7 ppm for 35 morning samples. The discrepancy is over an order of magnitude.

Figure 1 suggests that although the CH_4 and CO differences between the bag and can samples are minimal, the can TNMHC values seem significantly lower than those of the bag. All but one of the can TNMHC concentrations were lower than the lowest bag value. A more detailed examination of the bag and can differences is discussed by Seila (1979), with the conclusion that the can values are probably more representative of the true JSF concentrations, and that the bag samples were probably contaminated by permeation of fuel vapors during the flight from Houston to Research Triangle Park. The individual HC's which showed no significant bag/can difference were ethane, propane, α-pinene, and some of the aromatics. Most of the HC's were considerably higher

4

in the bags than in the cans, although the oxygenates (acetaldehyde, acetone, and propionaldehyde) were considerably lower in the bags.

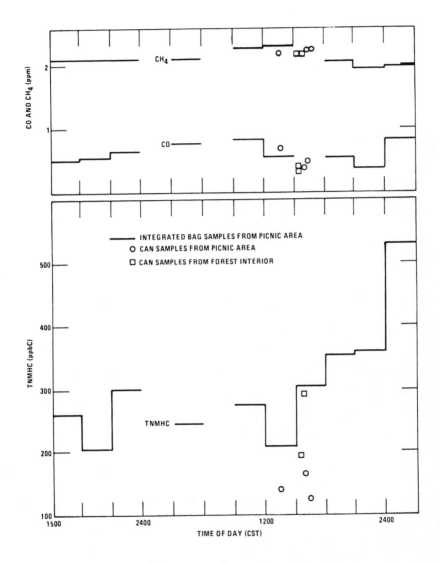

Figure 1. Jones State Forest TNMHC, CH$_4$, and CO concentration results for January 4-6, 1978.

The composition of Jones State Forest TNMHC based on the can data was paraffins (72%), aromatics (18%), olefins (6%), acetylene (2%), and vegetative HC (2%).

The composition of Houston commercial natural gas, computed from analysis of two samples is shown in Table I.

TABLE I. COMPOSITION OF HOUSTON COMMERCIAL NATURAL GAS

Compound	Percentage
Methane	90.19 ± 0.41
Ethane	5.16 ± 0.36
Propane	2.40 ± 0.12
n-Butane	0.78 ± 0.05
i-Butane	0.47 ± 0.04
n-Pentane	0.28 ± 0.03
i-Pentane	0.28 ± 0.02

SOURCES OF JONES STATE FOREST TNMHC

An analysis was performed based on the HC to acetylene (C_2H_2) ratios within JSF, the Houston ambient air, and two Houston tunnels to determine the sources of Jones State Forest TNMHC. If two assumptions are made: that vehicular emissions are the only significant source of C_2H_2 and that C_2H_2 is unreactive in the atmosphere, then C_2H_2 can be used to trace both auto exhaust and the Houston urban plume. The appropriate ratio of source HC to C_2H_2 multiplied by the JSF C_2H_2 concentration yields the HC concentration in JSF due to the particular source. Similar C_2H_2 ratioing techniques have been used by others in the past (Lonneman et al. 1974, Stephens and Burleson 1969, McMurry et al. 1975).

For this analysis to be valid, the wind during and somewhat prior to the sampling period at JSF had to be from the general direction of Houston, which is to the south. Fortunately, the wind was indeed from a southerly direction during the entire 3-day study. When the can samples were collected on January 5, the mean hourly average wind direction from 10:00 CST to 16:00 CST was $192° \pm 5°$ at the Texas Air Control Board's Aldine station. The mean hourly average wind speed during the same 8-h period was 21.4 ± 1.7 m/sec (Driscoll 1978).

The sample data used for the analysis of the HC to C_2H_2 ratios were obtained from two sources: from this report and from the Houston metropolitan area and tunnel (Baytown and Washburn) data reported by Lonneman et al. 1979. This earlier study conducted by Lonneman included two tunnel samples and 19 ambient samples distributed between Jacinto City to the north, Baytown to the east, Pasadena to the south, and the I-45/I-10 interchange to the west. Seven of the ambient samples were collected on a very stagnant day; their TNMHC values ranged from 2.9 ppm to 9.1 ppm.

The HC to C_2H_2 ratios of the most abundant compounds observed in JSF were determined for Houston ambient air, tunnel air, and JSF air. The ratios are plotted in bar graph form in Figure 2. The three bars under each group or individual compound represent, from left to right, the compound to acetylene ratio in Houston tunnel air (vehicular emissions), Houston metropolitan air, and JSF air. This bar graph provides a visual basis for explaining the qualitative relationships between the various ratios and the HC sources.

If diluted vehicular emissions were totally responsible for the HC's in JSF, the three lines under each compound would be of nearly equal height. Although the paraffins and aromatics do not show this relationship, the olefinic compounds do, suggesting that the predominant source of olefins in JSF is vehicular emissions. The slightly decreasing olefin/acetylene trend indicates that the olefins have reacted somewhat during transport to JSF.

Hydrocarbon sources other than vehicles would be revealed by lines (HC/C_2H_2 ratios) higher than the corresponding vehicular emissions ratios. The increased ratios of all of the paraffins and aromatics of Houston air relative to the tunnel indicate that there are sources of these compounds in Houston other than vehicular emissions. In addition, that the ratios of some of the paraffins (ethane, propane, isobutane) are higher in JSF air than in Houston air indicates that the sources of these paraffins are other than Houston or vehicular emissions. These sources must be located between the Houston and JSF sampling areas. Hereinafter, this location will be referred to as "north Houston" and defined as the area between I-610 and the JSF boundary.

The previous graphical analysis suggests three sources for Jones State Forest TNMHC: vehicles, Houston, and north Houston. However, a fourth source can also be added, JSF vegetation. The contribution of vegetative sources was determined by direct measurement of the natural HC's in the JSF samples. The contributions of each of the other three sources were determined by calculations based upon HC to C_2H_2 ratios. The Houston and north Houston sources were further subdivided into individual and grouped HC contributions. The Houston and north Houston categories do not include vehicular emission compounds from those areas; these compounds are from sources other than vehicles. The results of these calculations are shown in Table II; the methods of calculation and error estimation are described elsewhere (Seila 1979).

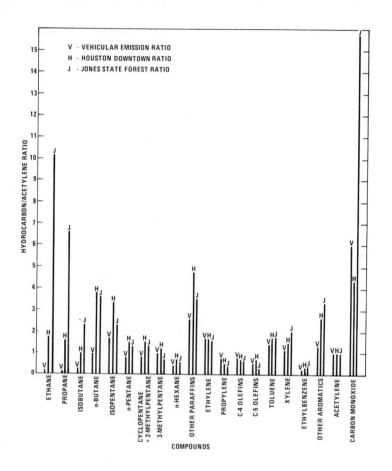

Figure 2.   Hydrocarbon-to-acetylene ratios of Houston vehicular emissions, Houston downtown air, and Jones State Forest air.

Vehicular emissions comprise 35% of Jones State Forest TNMHC. This percentage is only slightly higher than the vehicular contributions to C_2 to C_5 HC's determined by McMurray et al. (1975) and Westberg (1978) for several sites in the Houston metropolitan area itself.

TABLE II. SOURCES OF JONES STATE FOREST TNMHC

Sources	Compounds	Concentrations + Estimated Standard Error (ppbC)		Percent
		Individual	Total	
Vehicles			46.0 + 8.8	34.6
Houston*	n-Butane	7.7 + 1.9		
	Propane	4.0 + 1.0		
	Ethane	3.6 + 0.9		
	Isobutane	2.0 + 0.5		
	n-Pentane	2.0 + 0.7		
	Isopentane	1.4 + 1.2		
	Other paraffins	2.7 + 2.2		
	Aromatics	5.7 + 2.2		
	Total		29.1 + 9.7	21.9
North Houston*	Ethane	23.1 + 5.9		
	Propane	13.3 + 3.7		
	Isobutane	3.5 + 0.9		
	Aromatics	2.9 + 4.3		
	Total		42.8 + 8.8	32.2
Vegetative HC's			3.0 + 0.4	2.2
All			120.9 + 12.7	90.9
Measured TNMHC			133.0 + 15.4	100.0
Unspecified TNMHC			12.1 + 20.0	9.1

*Excluding vehicular emissions

The Houston metropolitan area HC's from sources other than vehicles are primarily light paraffins (ethane, propane, butane, pentane). Some heavier paraffins and aromatics were also present. The sources of these compounds could be refinery, industry, and/or natural gas. Although the light paraffins are compounds in natural gas, the percentage of the ethane and propane components of the Houston non-vehicular TNMHC of Jones State Forest is too low to represent much natural gas. If the assumption were made that all of the

9

non-vehicular ethane in JSF due to Houston (3.6 ppbC) were from natural gas, then the natural gas concentration in JSF due to Houston would be 3.6 divided by 0.57 = 6.3 ppbC, where 0.57 is the ratio of ethane to TNMHC in Houston natural gas. Of the NMHC's in Houston other than vehicular emissions, no more than 22% (6.3 divided by 29.1) can be due to natural gas emissions; the rest must be due to refinery or industry sources.

The non-vehicular HC's emitted between Houston and JSF (north Houston) are also primarily light HC's--ethane, propane, iosbutane--and some aromatics. In this case, the proportions of ethane and propane suggest that the natural gas component of non-vehicular TNMHC in north Houston is higher than in Houston air. Again, the assumption that natural gas is the source of all ethane leads to a computation that 95% of the non-vehicular TNMHC in JSF from north Houston is natural gas. The use of propane or isobutane for the same computation yields values over 100%. This is strong evidence of very significant natural gas emissions from the north Houston area.

The CO to C_2H_2 ratio in JSF was much higher than either the vehicular or Houston CO to C_2H_2 ratios, indicating sources of CO other than vehicles (e.g., industrial combustion). The HC measurements of Westberg et al. in 1976 at a site north of Houston showed "high aromatic content" (Westberg et al. 1978), which too suggests industrial activity. The concentration of ethane in JSF unaccounted for by dilution of Westberg's north site air was computed to be 15 ppbC. These facts suggest that the sources of light HC's in north Houston are probably industries which use natural gas for combustion. However, that geogenic oil and natural gas seeps exist along the east coast of Texas (Davis 1967) suggests that the possibility of some geogenic natural gas being present in JSF cannot be excluded.

Vegetative HC's

Jones State Forest is predominantly a loblolly pine forest, but some hardwoods, primarily oak and sweetgum are also present. Isoprene and α-pinene are the major emissions of hardwoods and loblolly pine, respectively (Dimitriades and Altshuller 1977). These are the natural HC's one would most expect to see at JSF. Lesser emissions one might expect are d-limonene, β-pinene, Δ-carene, and β-phellandrene (Zimmerman 1979).

The results of this study were no exception to previous findings by this and other laboratories that ambient concentrations of natural HC's were in the low ppb carbon range. Only α-pinene and isoprene were observed in any of the samples: isoprene in two and α-pinene in fourteen (all but one). The two isoprene measurements were 0.4 ppbC and 1.2 ppbC; the mean α-pinene concentration (from a range of less than 0.1 ppbC to 7.7 ppbC) was 3.6 ppbC $\pm$ 0.5 ppbC. As Table II shows, the natural HC contributed to Jones State Forest TNMHC was only 2%.

The diurnal variation of α-pinene compared to that of ethane and the wind speed is plotted in Figure 3. The plot shows that the α-pinene concentration increases rather sharply at sunset as the nocturnal inversion sets in marked by lowered wind speed. The concentration begins to decrease during the night as the mixing height above ground rises and decreases further during the day after the inversion has broken and wind speed has increased. This diurnal behavior is precisely that which one would expect from a local source such as the pine trees in JSF. In contrast, the ethane concentration shows generally the opposite behavior of α-pinene. Its concentration falls as the nocturnal inversion begins, but rises during the day when there is considerable mixing from above. This behavior indicates that the source of ethane is not local, but exists outside of JSF. The peculiar rise of ethane and drop of α-pinene during the last 3-h sampling period was probably due to a breakup of the nocturnal inversion by rain in the night. The diurnal behavior of α-pinene indicates that meteorological conditions influence the concentration of vegetative HC's. Daytime meteorological conditions of full sun and good mixing render very remote the possibility of vegetative HC concentrations rising to levels of photochemical significance.

CONCLUSIONS

The conclusions herein concern the ambient air concentrations, composition, and sources of HC's at JSF. Because of the limited nature of the sampling program, the conclusions are valid only for the period of the study when the wind was from a southerly direction.

The TNMHC concentrations at JSF ranged from 0.1 ppm to 0.5 ppm and consist primarily of paraffins. The TNMHC composition during this study was 72% paraffin, 18% aromatic, 6% olefin, 2% acetylene, and 2% vegetative.

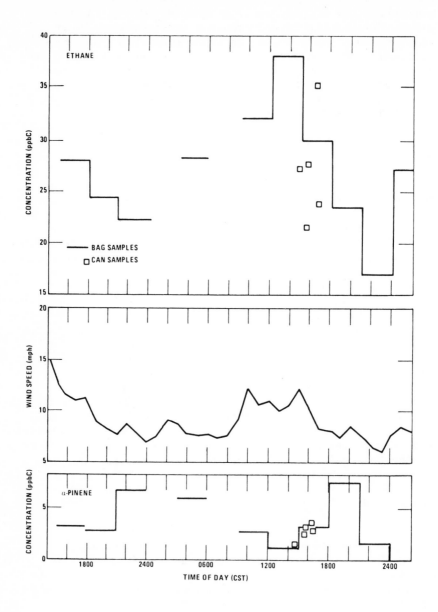

Figure 3. Diurnal variation of α-pinene, ethane, and wind
speed at Jones State Forest, January 4-6, 1978.

The vast majority of TNMHC in the air of JSF, 89% in this study, were transported there from outside sources. The sources can be divided into two main categories: vehicular and non-vehicular. In this study, vehicular sources accounted for 35% and non-vehicular sources accounted for 54% of Jones State Forest TNMHC. Of the non-vehicular HC's, 41% were due to Houston sources, and 59% were due to sources north of Houston. Refinery and natural gas emissions appear to account for most of the non-vehicular HC's.

The vegetative HC concentrations at JSF were low; they were less than 10 ppbC and represented only 2% of the TNMHC burden.

REFERENCES

Coffey P. E. and H. Westberg. 1977. Part IV. The issue of natural organic emissions. In: International Conference on Oxidants, 1976: Analysis of Evidence and Viewpoints. EPA-600/3-77-116. U.S. Environmental Protection Agency, Research Triangle Park, North Carolina.

Davis, J. B. 1967. Petroleum Microbiology. Elsevier Publishing Company, New York, New York, p. 98.

Dimitriades, B. and A. P. Altshuller. 1977. Part I. Definition of key issues. In: International Conference on Oxidant Problems: Analysis of the Evidence/Viewpoints Presented. J. Air Pollution Control Assoc. 27(4):299.

Dimitriades, B. and A. P. Altshuller. 1978. Part II. Evidence/viewpoints on key issues. In: International Conference on Oxidant Problems: Analysis of the Evidence/Viewpoints Presented. J. Air Pollution Control Assoc. 28(3):207.

Driscoll, T. 1978. Texas Air Control Board, Houston, Texas. Private communication, February.

Lonneman W. A., S. L. Kopczynski, P. E. Darley, and F. D. Sutterfield. 1974. Hydrocarbon composition of urban air pollution. Environ. Sci. Technol. 8(3):229.

Lonneman, W. A., G. R. Namie, and J. J. Bufalini. 1979. Hydrocarbons in Houston Air. EPA-600/3-79-018. U.S. Environmental Protection Agency, Research Triangle Park, North Carolina.

Mayrsohn, H. and J. H. Crabtree. 1976. Source reconciliation of atmospheric hydrocarbons. Atmos. Environ. 10(2):137.

McMurray, J. R., R. E. Flannery, L. H. Fowler, and D. J. Johnson. 1975. Ambient sampling for stationary and mobile source hydrocarbons in Houston, Texas. Presentation: Paper 75-45.5, 68th Annual Meeting Air Pollution Control Association, June.

Seila, R. L. 1979. Non-urban Hydrocarbon Concentrations in Ambient Air North of Houston, Texas. EPA-600/3-79-010. U.S. Environmental Protection Agency, Research Triangle Park, North Carolina.

Seila, R. L., W. A. Lonneman, and S. A. Meeks. 1976. Evaluation of polyvinyl fluoride as a container material for air pollution samples. J. Environ. Sci. Health-Environ. Sci. Eng. A11(2):121.

Stephens, R. E. and F. R. Burleson. 1969. Distribution of light hydrocarbons in ambient air. J. Air Pollution Control Assoc. 19(12):929.

Westberg, H., K. Allwine, and E. Robinson. 1978. Measurement of Light Hydrocarbons and Oxidant Transport-- Houston Area 1976. EPA-600/3-78-062. U.S. Environmental Protection Agency, Research Triangle Park, North Carolina.

Whitehead, L. and R. K. Sievers. 1977. Background hydrocarbon levels in east Texas. Presentation: Paper 21a American Institute of Chemical Engineers 83rd National Meeting, Houston, Texas, March.

Zimmerman, P. R. 1979. Testing of Hydrocarbon Emissions from Vegetation, Leaf Litter, and Aquatic Surfaces, and Development of a Methodology for Estimating Emission Rates from Foliage. EPA-450/4-79-004. U.S. Environmental Protection Agency, Research Triangle Park, North Carolina.

DISCUSSION OF PRESENTATION

RASMUSSEN: You would say the ethane, the propane, and the isobutane are from the forest?

SEILA: No, I would say they're from the area between Jones State Forest and Houston. I'm saying that ethane and propane are from sources other than vehicular emissions.

These sources are due to Houston, the Houston plume itself, and also to the region between Houston and JSF.

RASMUSSEN: What is the paraffin source that gives this exceedingly high ethane, or propane and isobutane? Have you identified it?

SEILA: I don't know. I suspect, possibly natural gas emissions. In Figure 2 we show the aromatic HC's and CO. Again, we see sources other than vehicular emissions due to Houston, and also sources between Houston and JSF--not as exaggerated in some cases, however, as the paraffins.

Also, there appear to be sources of CO between Houston and JSF.

RASMUSSEN: How did you determine the vegetation was 2%?

SEILA: That was just our measurement. That's not a calculation.

RASMUSSEN: What was in the ambient air? Specifically, what compound?

SEILA: Isoprene and α-pinene.

RASMUSSEN: That's completely fallacious. I mean, there are more compounds released by vegetation than just isoprene.

SEILA: That's all we were able to measure, though.

RASMUSSEN: Yes, but Pat Zimmerman's data show again and again that if you look at what the vegetation is giving off, only around 28% in the daylight in the Tampa area was attributable to isoprene, and another 19% to the terpenes; at most it comes up to about 47%.

Half of the organic material, the nonmethane type, is of some other indiscriminate HC species.

STEVENS: It's quite a bit different in Houston.

RASMUSSEN: I don't think that's a valid approach.

SEILA: Well, I'm saying that if we assume that maybe we weren't able to measure some of the vegetative HC's, still we're only unable to account for 9%.

RASMUSSEN: I'll go along with that.

15

SEILA: If you want to assume that the 9% is due to vegetative sources, fine. Add that onto the 2% and you have 11%, which still isn't very much.

RASMUSSEN: I think that's sloppy. You go to all of this detail about what comes from vehicles, what comes from Houston, what comes from north Houston, then you put down "Vegetation, TNMHC's." What you're really saying, under "Vegetation," is isoprene and the terpenes.

SEILA: Yes, I'm adding the two together.

RASMUSSEN: But you still haven't put in there the other component. I think you ought to double that. I mean, what are we arguing about? Two percent versus four percent may be a small difference, but I've seen this occur again and again. Everybody takes that to mean that the only thing plants give off is isoprene and the terpenes; when, in fact, they give off a whole host of other organics.

SEILA: I agree. I'm not suggesting that these are the only vegetative HC's. There are others. But I can't discriminate between those other vegetative HC's. I don't know what they are.

RASMUSSEN: I agree with that; it's very difficult. But there should be a caveat--

SEILA: So you're saying I should have named that something else? I should have said what it was specifically.

BUFALINI: Would it help if he just put down isoprene and α-pinene there?

RASMUSSEN: Yes, it would be helpful to include a footnote saying that the specific species could be identified, recognizing that there is something else. There is a whole host of other things that come off vegetation.

BUFALINI: I agree with what you're saying. But even if you bend over backwards, the best you're going to do is that unspecified 9%. I would strongly doubt even half of that 9% is due to vegetative sources. Let's not forget the basic reason for doing the study was to prove (as Bob has proved decisively) that the Whitehead-Sievers work was incorrect.

RASMUSSEN: I'll go along with you there. I'm not trying to say that isoprene or α-pinene plus the other HC's from vegetative sources are a dominant source of HC in the analysis made at these sites. I do think, however, that it's time to make some very specific points. There is more given off than isoprene and α-pinene. When you lump everything together as TNMHC, and then isolate it and forget to identify it, you always end up talking about only isoprene and α-pinene.

In the paper I originally published on my thesis, with Dr. Went, we only claimed to be measuring and plotting the total organic volatiles at the sites. We illustrated that we could measure isoprene and α-pinene in these peaks. Everybody since that time has taken this to mean that all we were plotting was α-pinene or isoprene, and that was not the case, as is made clear in my paper. I think this assumption has been propagated too long.

BUFALINI: Well, again, you're right. But when Bob [Seila] goes through his chromatograms, the only thing he can identify is isoprene and α-pinene. I would suspect, as Katz found, that there are a number of unidentified peaks. But the problem is, when you go through and calculate how much carbon there is, there's so little that for all practical purposes it would be impossible to construct, other than to say that it was just isoprene and α-pinene. I believe that's the best he could do, because that's all he could identify.

RASMUSSEN: I accept that. I think that's the point I'm trying to get across. The other point I was hoping to make is that there should be some explanation of why in the vegetative emissions studies, they're constantly coming up with aromatic materials, as well as the olefins and the paraffins. I think these should be identified a bit more specifically.

LUDLUM: I think he's hit the nub of the whole problem; you identify terpenoid compounds and isoprene, but not the third category that was on his slide yesterday.

SEILA: Yes, we recognize that. But, if there are aromatics present, we can't necessarily differentiate between those and the anthropogenic aromatic emissions from cars and elsewhere.

STEVENS: I think that 9 ppbC was unidentified.

SEILA: Right. Unaccounted for.

STEVENS: Unaccounted for. This 9 ppbC is a combination of both natural emissions and photochemical products of both automotive exhaust and other sources. The array of 50, 60 peaks we saw in the chromatograms are from different biogenic sources, and each peak represents only a very small amount of total carbon. I would imagine you can see hundreds of different compounds, but are they significant to the NMHC burden as it exists in the atmosphere?

I think your data clearly show that they are <u>not</u> significant. They represent a minor <u>portion</u>, and any effective control strategy for HC's cannot be addressed to the minor constituents.

BUFALINI: Could I interrupt for a moment? We don't want to get too far behind. I like the discussion, but this could go on for another half an hour.

DIMITRIADES: Okay, but this is an issue, and I think we need to discuss it.

RASMUSSEN: This is the issue we came to discuss.

BUFALINI: I don't disagree with you, but we're talking about a rather small quantity. I think everyone here would agree, unless Ken [Dr. Ludlum] would disagree, that we're still talking about extremely low concentrations. I agree that they should be identified, but he has to bend over backward to ascribe that 9% to natural HC's.

SEILA: There's a certain amount of analysis error here, too. I've tried to calculate the errors.

RASMUSSEN: Just one last point. I'm not saying that one has to ascribe the 9% to natural HC's. I'm raising the question of the vehicular emissions from the Houston and non-Houston areas. How much of the paraffin and aromatic material now claimed to come from vehicles in the non-Houston area might be a component of what had been measured under situations where he was not compromised by any anthropogenic emissions? There is a small percentage in those numbers, too.

SEILA: I think that's accountable to the error, because that's a number calculated from our data. That number is based upon the automobile exhaust and our concentrations in the Houston area.

RASMUSSEN: Well, I think we are quibbling over small percent differences.

SEILA: Yes, we are.

RASMUSSEN: I think one ought to quibble over the point that the aromatics and paraffin material coming off the vegetation are included in your 46 ppbC, and your 29 ppbC, and your 43

ppbC attributed to these other purely anthropogenic sources.
In my best judgment, I would say that at most maybe the
vegetation was--

SEILA: I don't agree with that.

STEVENS: That's not true, Rei [Dr. Rasmussen]. He handles
that through the ratio of the acetylene.

SEILA: Right.

STEVENS: I mean, you don't understand the purpose of what
he's doing.

SEILA: I've traced the automobile exhaust and the plume.

DIMITRIADES: Can't you resolve this by taking the ratios of
those unidentified peaks to acetylene, and seeing whether or
not they are comparable to the ratios from the exhaust
chromatograms?

SEILA: We're not really talking about unidentified peaks.
The unaccounted for part is not necessarily unidentified
peaks. It just indicates that the analysis method could not
completely account for the total amount that we analyzed.

DIMITRIADES: Oh, they were identified, but you couldn't
account for it in terms of vehicular contributions.

SEILA: It doesn't balance, in other words.

STEVENS: Which includes the uncertainty in NMHC
measurements, and uncertainties in gas chromatography.

SEILA: It includes a lot of uncertainties, for example, the
uncertainties of analysis and calculations - the variance.

DIMITRIADES: This is certainly something to be looked at. I
don't think this is trivial - the difference between 2% and
4%. It may be trivial with respect to the problems in urban
areas, but when it comes to assessing the role of the natural
organics in regions, etc., then it's something to give some
further attention.

BUFALINI: I don't agree with that at all. Let's not forget
that this was within the canopy.

SEILA: Beneath the canopy.

BUFALINI: Beneath the canopy, okay. Whenever you go into a
forested area, you expect to observe a lot more HC. If

you're going to say that 4% is significant and 2% is not, well, that's interesting, but what would it be 5 km downwind?

DIMITRIADES: I appreciate that. It isn't the percent that interests me, it's the amount. If you are missing 2%, obviously some of those peaks (presuming your're missing natural organics) are natural organics, and you don't know how much right now. I guess that's what I'm getting at.

SEILA: Yes. Part of the 9% may be natural HC's.

Briefly, this is a summary of the natural HC observations made during the study. My final graph shows a diurnal variation of methane compared to α-pinene. What we noticed here was that at night the α-pinene values increased, and in the daytime they decreased. The opposite phenomenon was observed for the ethane. This suggests to me that when the inversion sets in at night, the air in the forest is separated from the air being transported from Houston. This causes a decrease in our ethane and an increase in the concentration of the α-pinene in the forest.

During the day, when you have good mixing height, the ethane is being transported from Houston and vicinity. It mixes in the forest and you have more air dilution. Thus, the α-pinene concentrations go down and the ethane concentrations go up.

This is just a summary of the results from the calculations. I want to point out that the vegetative HC's we measured, isoprene and α-pinene, were only 2% of the NMHC in Jones State Forest.

RASMUSSEN: Bob, don't get us wrong. I think you did a terrific job.

WALKER: As I understand it, these data were taken in January?

BUFALINI: Yes, however, Richard Sievers conducted a study in which he collected data througout the year, and he also claims to have seen extremely high concentrations.

WALKER: Well, I just wanted to note that those trees in particular won't have leaves on them in January. The trees do lose their leaves for a couple of months in Houston.

BUFALINI: We did ask Sievers about that, and he claimed that he had done the studies throughout the year and saw high concentrations during the winter also.

WALKER: Well, there's no question about that. He refuted Sievers' work. But the point is that in June you might see more, particularly isoprene, because the vegetation is more active.

BUFALINI: Probably.

SEILA: Yes.

WALKER: Another point I would like to inquire about is that you're assigning a certain distribution of HC's to "Houston," as distinct from "vehicles"? Is that my understanding?

SEILA: Yes.

WALKER: During the Houston Area Oxidant Study (HAOS) work, we did a lot of sampling. The bulk of the Houston sampling we saw looked a lot like vehicles. Now, I don't say all of it was, but it wasn't fixed. It varied widely. So if you only had a few samples, they may not be totally typical.

SEILA: Well, you have to consider that the analysis calculates the contribution in JSF, not the contribution in downtown Houston. I assumed dilution and I—

WALKER: Yes, but you assumed a distribution, a fixed distribution for "Houston" sources of vehicles. Is that right?

SEILA: Yes, we assumed the distribution; we took the two tunnel samples we had, and from those calculated the various hydrocarbon-to-acetylene ratios.

WALKER: Were those for vehicles?

SEILA: Yes.

WALKER: What did you use, solving for Houston?

SEILA: We used 18 ambient samples that had been collected in the Houston area between Pasadena and downtown.

WALKER: I would think that the 18 varied rather widely. Thus an average really wouldn't tell you much.

RASMUSSEN: Are those ground samples in Houston, or air samples?

SEILA: Ground samples. There was a lot of variance in the actual concentrations, but there was less variance in the ratios, which is the important thing.

21

WALKER: But the ratios probably didn't differ widely from vehicular ratios, did they?

SEILA: There was quite a bit of difference. For some particular compounds, I showed, there wasn't much difference. For example, there wasn't much difference for the olefins; the point of the bar graphs was that there wasn't much difference in the olefinic ratios. However, for certain compounds, particularly ethane and propane, there was a lot of difference in the ratios.

DENYSZYN: I would like to make a couple of points. The Baton Rouge to Houston area is the largest producing area of acetylene in the U.S. When Research Triangle Institute (RTI) did sampling in the area in 1976 or 1977, this was something that had to be realized, that the area was heavily influenced by acetylene plants. So, be careful with your acetylene ratio.

The other point I think should be considered even more seriously is that the vehicular emissions you're using to evaluate your vehicular ratios are from 1977 samples. Remember that the intrusion of the catalytic converter is strongly affecting that particular ratio, acetylene, and especially the paraffinic ratios. So, you should be careful when using those particular samples, which are much older, to introduce these ratios.

SEILA: I agree. Although this is a method for evaluating or partitioning the anthropogenic HC's from the natural HC's, it would of course be much better if all the samples were collected at the same time.

DENYSZYN: The fallacy is that you're making a broad conclusion about the overall aspects of the city with data which are not what I would consider of the highest quality; you've got 1977 data.

SEILA: Yes, I know. I agree that it's a problem with this particular analysis. The problem could be overcome, if someone wanted to do the same thing elsewhere. They'd need to collect all their samples at the same time.

WALKER: I would like to differ with the last observer. There are only two acetylene plants left in Houston, if any. Their total losses would be negligible in the total scheme of things down there. I can't imagine you could detect acetylene at any range from the immediate plants. I think the losses from welding cylinders would have a far greater effect on the area, and there is a lot of welding done in Houston.

BUFALINI: I don't think that acetylene is necessarily the only indicator. I think you can strategically ratio these HC's, and--

DENYSZYN: I'm perfectly aware of that.

BUFALINI: I think your point is well made, but I don't really believe that acetylene was a significant source.

DENYSZYN: I'm not saying it's a significant source, but I think it's a contributing source for that area.

BUFALINI: Okay, you're probably right.

SEILA: Let me say that I calculated the errors of the ratios. If you notice, some of the errors are 20% or 25%. I wanted to include that so you wouldn't think this was a definitive analysis of Houston. There is a certain amount of error in these calculations.

LUDLUM: I have a question concerning the ethane, propane, and isobutane ratios. These ratios were very different, but you attributed them to north Houston or natural gas seep. I just want to point out that there could have been a gas seep at your feet, while you were measuring it. It didn't have to come from north Houston.

SEILA: I arbitrarily established north Houston as the area between the belt-line (I've forgotten the road number) and JSF. The presentation had to be short, so I really didn't get into a definition of some of the terms.

LUDLUM: I guess my real question is, couldn't you say something about the expected ratios of isobutane to ethane in natural gas, and see if those residual ratios fit?

SEILA: I did that in my report [EPA-600/3-79-010].

WALKER: I just want to note that the HAOS study showed excess paraffins in a lot of its analyses. These paraffins were not just on the north side; a lot of them were on the south side.

GRAEDEL: Would your detection technique pick up oxygenated species like camphor?

BUFALINI: It would depend on whether or not it would come up to about C-12. I don't know for sure.

LONNEMAN: I don't know.

BUFALINI: It would probably be borderline, at best.

GRAEDEL: Would it be possible that there could be a moderate concentration, then?

SEILA: I think there could possibly be some oxygenated compounds that we did not pick up. But, I don't think the concentration is very high.

GRAEDEL: Well, why do you say that?

SEILA: Well, we can measure acetone, acetaldehyde, and propionaldehyde.

GRAEDEL: I'm trying to contrast here an oxygenate that you'd expect from the chemistry of automotive emissions to one that you might get naturally from vegetation. It's not obvious to me that your argument on acetone applies to camphor.

SEILA: No, it doesn't necessarily apply to camphor. We can pick up camphor; whereas, some of the other heavier ones, I'm not so sure we can pick up.

11. BIOGENIC HYDROCARBON MEASUREMENTS

Hal H. Westberg. Department of Chemical Engineering,
Washington State University, Pullman, Washington 99164

ABSTRACT

Research techniques are described that were developed to
establish the identity of organic compounds emitted by
vegetation. The validity of various procedures used for
collection, concentration, and analysis of natural
hydrocarbons in ambient environments is also discussed.

Stainless steel canisters appear to provide an
acceptable collection medium for terpenic hydrocarbons, and
cryogenic and/or solid adsorbent concentration procedures do
not appear to compromise sample integrity. Qualitative and
quantitative determinations were made using gas
chromatographic and gas chromatographic/mass spectrometric
techniques.

Results from field measurement programs have shown that
ambient terpene concentrations are generally very low (< 10
ppbC). Our emissions studies have shown that terpenes
normally account for 80 to 85% of the measurable hydrocarbons
from natural vegetation, and that oxygenated species appear
to comprise a very minor part of vegetative emissions.

INTRODUCTION

This paper contains a brief review of terpene oxidation
studies that have been carried out at Washington State
University (WSU). Major portions of the paper discuss
ambient measurement of terpenes. Also mentioned are factors

25

that would appear to affect terpene emissions from vegetation.

Kinetic studies on terpene photooxidation and ozonolysis reactions were initiated at WSU in the fall of 1971. Prior to that time, a few studies had been performed that indicated terpenes were photochemically reactive; however, no systematic study of terpene reactivities had been undertaken.

Table I shows results from our terpene reactivity studies. Of note is that all terpenes are photochemically more reactive than isobutene. As would be expected for olefinic types of hydrocarbons (HC's), the reactivity rates were very rapid when these terpenes and other compounds were combined with ozone (O_3).

TABLE I. RELATIVE REACTION RATE (PHOTOOXIDATION)

Compound	k (Relative)
Isobutene	1.0 (k = .84 x 10^{-4} sec^{-1})
α–Pinene	1.6
β–Pinene	1.3
Δ^3–Carene	1.7
Limonene	2.9
Myrcene	4.6
Terpinolene	13.0
Isoprene	1.6

Product studies were conducted concurrently with the kinetic studies. Terpenes were found to behave in a manner analogous to anthropogenic HC's when exposed to sunlight in the presence of nitrogen oxides (NO_x).

Figure 1 illustrates data obtained from terpene photooxidation studies, showing typical results of concentration versus reaction time. Note that both the terpene and nitric oxide (NO) concentrations decrease rapidly and that O_3 and nitrogen dioxide (NO_2) were produced. Further, the amount of O_3 formed was dependent on the absolute concentration of the reactants, and also on the terpene-to-NO_x ratio. Optimum O_3 production occurred at a ratio of about 15.

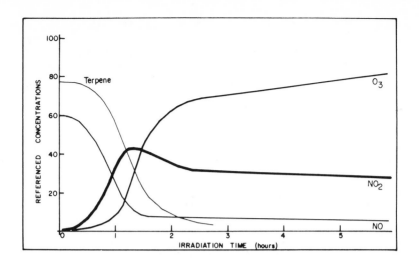

Figure 1. Typical graph of concentration
versus reaction time.

The question of aerosol production was also pertinent,
due to early reports by Went that terpene oxidation is
probably responsible for the haze observed over forested
regions in summer (Went 1960). The fact that a large amount
of light-scattering aerosol is produced when terpenes are
photooxidized or combined with O_3 was verified. From a
quantitative standpoint, our best estimates were that roughly
50 to 75% of the carbon went into the aerosol phase, and that
the efficiency of O_3 production from terpenes was quite low.
In the smog chamber, roughly 20 ppb carbon were required to
produce 1 ppb O_3. Hence, basic terpene behavior in the
atmosphere should be the same as that of olefinic HC's from
combustion sources.

The next question to pursue was whether terpene
concentrations are sufficiently high in rural areas to
adversely affect air quality. Early reports appearing in the
literature, as well as our initial atmospheric measurements,
revealed very low terpene concentrations. Rasmussen and Went
(1965) reported 10 ppb in a forested area of the Midwest,
while our studies in forested areas of Washington, Oregon,
and California during the mid 1970's showed essentially no
evidence of terpenes. Our results were somewhat perplexing
since laboratory studies had verified that terpenes were
emitted by vegetation, and emission estimates based on those
studies indicated that natural HC's should be detectable in
forest environments.

But are the terpenes really absent, or was the analytical methodology for measuring the terpenes faulty? In order to answer this question, the analytical methods have been examined in detail. The normal procedure involves three steps: collection of an ambient sample, concentration of the organics in a portion of the sample, and the actual qualitative and quantitative determination.

EXPERIMENTAL

Qualitative and Quantitative Determination

The qualitative and quantitative determination is usually achieved using gas chromatography (GC). Fortunately, GC column technology has advanced quite rapidly during the last ten years; through the use of currently-available glass capillary columns, the terpenes can be separated from other HC species. Figure 2 shows a chromatogram obtained from a Ponderosa pine emission sample collected in a Teflon bag, using the enclosure method developed by Pat Zimmerman (1979). (Only the portion of the chromatogram containing the monoterpenes is included.) The main feature to note is that the individual terpenes are well resolved; also present is the oxygenated HC methyl chavicol, which consists of a benzene ring with a methoxy and a propenyl group in the 1,4 positions respectively. The type of chromatogram depicted in Figure 2 was obtained with a 30-m, SE-30, glass capillary column. A subambient temperature programmed analysis was employed, starting at $-60°C$ and rising to $80°C$.

Figure 3 shows an entire chromatogram: Peak #1 is isoprene; #2 is α-pinene; #3 is camphene; #4 is β-pinene; #5 IS myrcene; #6 is Δ-carene; and #7 is limonene. A question often asked is what percentage of the HC's in these emission samples is really terpenes? From Figure 3, the terpenes are by far the predominant species; in the vegetative emission samples studies at WSU, terpenes generally make up 80 to 85% of the HC total.

The chromatogram shown in Figure 4 contains many peaks. Of note is that terpenes can be resolved in a mixture containing anthropogenic-type HC's. This sample was collected in a parking lot and most of the HC's related to vehicle exhaust. The canister was spiked with isoprene, α-pinene, β-pinene, and limonene, just to show the relationship between their retention times and those of the

commonly-encountered anthropogenic HC's. Isoprene is fairly
well resolved, as are the monoterpenes, which have longer
retention times than toluene, ethylbenzene, and the xylenes.
Hence, the GC method appears to be acceptable, and would
reveal the presence of terpenes in ambient atmospheres, if
the collection and concentration techniques are valid.

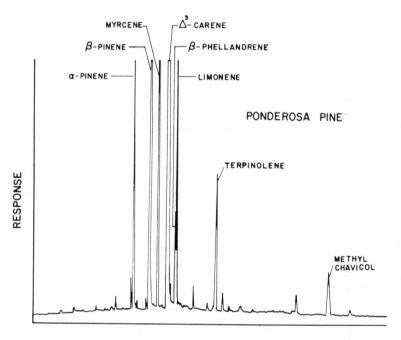

Figure 2.   Partial chromatogram of sample collected
            by Teflon-bag enclosure method.

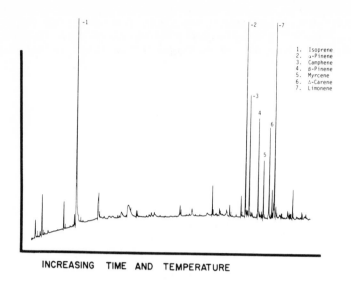

Figure 3. Entire chromatogram of a Ponderosa pine sample.

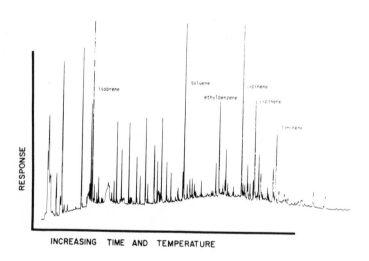

Figure 4. Chromatogram showing resolution of terpenes in a mixture containing anthropogenic-type HC's.

Concentration and Collection Steps

The concentration and collection steps were also examined. The chromatogram in Figure 4 was obtained using a cryogenic concentration procedure in which a small volume of air is passed through a freeze-out trap attached directly to the GC. The trap is packed with glass beads or glass wool, and submerged in liquid oxygen. The oxygen and most of the nitrogen in the air sample pass through the trap, while the organics are retained; the contents of the trap are then transferred to the GC column.

With the proficiencies of the concentration and chromatography steps verified, the remaining step to check was that sample integrity was maintained between the time of collection and GC analysis. This gap of time can vary from a few minutes, if the GC system is at hand, to several days, if samples must be shipped across country. The stability of a number of HC's in various collection containers has been examined. The superiority of the stainless steel canister was apparent from the start, and these containers have been studied most specifically in relation to the terpenes. Table II gives the results of HC storage tests in stainless steel canisters, obtained from an ambient air sample artificially spiked with terpenes.

TABLE II. HC STABILITY IN STAINLESS STEEL CANISTERS

Compound	C_i ($\mu g/m^3$)	C_7 ($\mu g/m^3$)	Δ (%)	C_{14} ($\mu g/m^3$)	Δ (%)
Isoprene	26.4	25.4	−4	23.9	−10
α-Pinene	13.1	11.9	−9	9.6	−27
β-Pinene	9.4	7.9	−16	5.6	−40
Limonene	8.4	6.1	−27	1.0	−88
2-Methylpentane	6.0	5.9	−2	5.7	−5
2-Methylhexane	7.5	7.0	−7	6.9	−8
Benzene	6.7	6.2	−7	5.9	−12
Toluene	14.8	14.3	−3	13.7	−7
m- & p-xylene	10.1	10.1	0	9.7	−4

The first column shows the initial concentrations immediately after the canister was filled. The concentration 7 days later is indicated by C_7; the concentration after 14 days is indicated by C_{14}. The percent change after 7 days and 14 days is indicated by Δ. If the nonterpene-type HC's are studied, the concentration change over this two week period in the canister is generally very small. The reproducibility of a series of injections is estimated at about 10%. Therefore, essentially no loss of the nonterpenes occurs over a 2-week period. The terpenes do disappear more rapidly than the paraffinic and the aromatic type of HC's; however, over a period of 1 week, the losses are small. Certainly, if the analysis is completed within a few days after the time of sample collection, terpene storage losses will be minimal.

Verification of Sample Integrity in Canisters

A number of other experiments were conducted to verify sample integrity in canisters. For example, two canisters were filled simultaneously, then NO was added immediately to one in an amount that lowered the O_3 concentration to zero. The NO-spiked containers did not exhibit enhanced terpene levels when contents of the two canisters were analyzed. Probably the most supportive experiments regarding sampling integrity were those in which forest air was collected simultaneously in stainless steel canisters and on the solid adsorbent Tenax. In this case, two completely different collection/concentration schemes were utilized, resulting in nearly identical terpene concentrations.

Analytical results that demonstrate the comparability of Tenax and canister collections are shown in Table III. The samples were collected during November 1976 in a northern Idaho forest. Two canisters and two Tenax samples were collected on the 16th, and two canisters and one Tenax sample were collected on the 19th. The terpene concentrations obtained by the two methods are very similar. From these experiments, the analytical methodology appears to be good, and faulty sampling procedures do not seem to be the cause for the low ambient terpene concentrations. Total terpene concentrations at our northern Idaho site, which is in a forest of mixed pine and fir, have varied between about 50 ppt and 1500 ppt, with the average being about 200 ppt (2 ppbC).

TABLE III.   TENAX AND CANISTER COLLECTION

Date	α-Pinene (ppt)	β-Pinene (ppt)	Δ^3-Carene (ppt)	Limonene (ppt)
Nov. 16, 1976				
canister	30	40	<10	<10
canister	40	40	<10	<10
Tenax	50	50	<10	<10
Tenax	25	40	<10	<10
Nov. 19, 1976				
canister	120	200	30	20
canister	110	200	30	10
Tenax	130	190	40	10

Mass Spectrometric Analysis

A specialized mass spectrometry (MS) technique should be mentioned that was used to measure terpenes at these low concentrations. Since the flow rate of a carrier gas through a capillary GC column is very low, the entire effluent can be removed from the column and put into the ion source of a MS, without the necessity of a jet separator or some other type of GC/MS interfacing device. The mass spectrometer can thus directly substitute for the flame ionization detector (FID). Maximum sensitivity is then achieved by utilizing the single-ion monitoring capabilities of the MS.

Figure 5 illustrates a single-ion chromatogram obtained from a forest sample (simply a graph of response vs. increasing time and temperature). The MS is set on a specific ion known to be most intense for isoprene, or any species of interest. The ion being monitored is changed throughout the run to correspond to the retention times of various HC's. For example, isoprene is monitored at m/e 68, but after the isoprene peak comes through, the selected ion is changed to 78 in order to observe benzene. Toluene is next, at m/e 91, followed by the terpenes at mass 93. This method provides a highly sensitive, very selective, means of identifying terpenes.

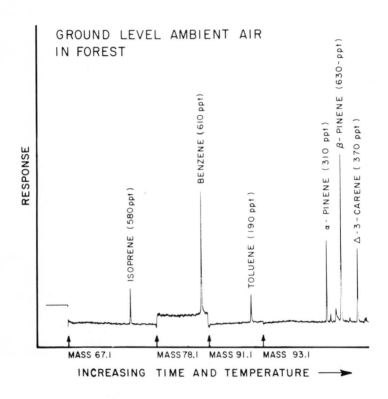

Figure 5. Single-ion chromatogram from northern Idaho
           forest sample (ambient air at ground level).

Figure 6 shows a single-ion chromatogram from a sample
that was collected outside the forest canopy. Although the
terpenes are no longer present, the same approximate levels
of benzene and toluene occur. Terpenes are generally not
detected outside the forest canopy.

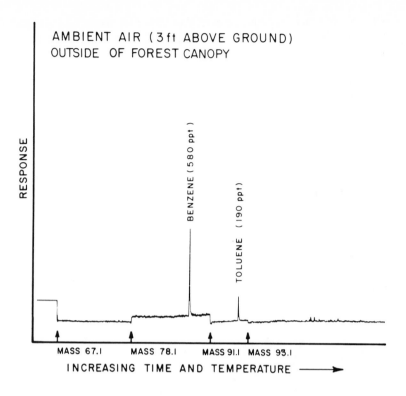

Figure 6. Single-ion chromatogram of sample collected
            outside the forest canopy (ambient air at 3 ft
            above ground level).

ADDITIONAL RESEARCH

     The considerable amount of natural emissions work at
WSU, initiated by Zimmerman and Rasmussen, has continued
since their departure. One area that has been studied is the
monthly variation in emission rates from two trees over the
period of about a year. Emission samples were collected
using the Zimmerman enclosure method for each month of a
14-month period (which spanned all four seasons). The two
forest species monitored were Ponderosa pine and red oak.
Figure 7 shows a plot of the pine emission rate in micrograms
per gram-hour for each month between May 1978 and June 1979.
In general, the emissions are fairly low, and the highest
rate of emissions was observed in late springtime. The first
point in Figure 7 (May 1978) is considerably off the top of
the graph, indicating a very high emission rate. Some
question exists as to whether that absolute value is actually

correct, but higher emissions during the May-June period from pine have been observed. A slight increase in emissions apparently occurs in the fall as well.

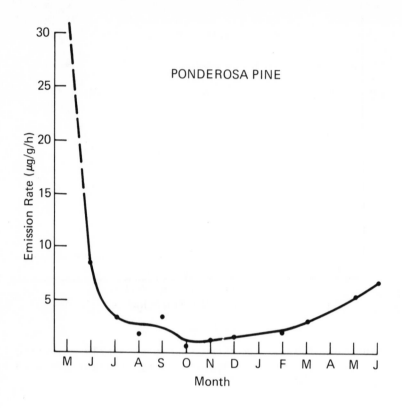

Figure 7. Emission rate for Ponderosa pine.

    Changes in HC composition of the emissions from Ponderosa pine were also observed. In the period when the volatile emissions were high (springtime), the largest component by far was Δ-carene. As the summer progressed, α- and β-pinene became more prominent. The literature indicates that Δ-carene is the primary component present in the oleoresin of northern Idaho Ponderosa pine. Volatile emissions from the oleoresin would seem to play a major role in spring, while in summer, the needle oil containing mostly β-pinene becomes more important.

With the red oak, a peak in emissions definitely occurs
during the summertime (illustrated in Figure 8) which, since
biomass is greater and temperatures are higher, is not
unexpected. These factors are known to affect isoprene
emission rates and thus significantly affect emissions from
red oaks, since isoprene represents 80 to 85% of the total HC
emission from this species. A good correlation exists
between emission rate and temperature with the isoprene-
emitting hardwood species, as Figure 9 illustrates from data
collected in the summer of 1979 in Pennsylvania. The slope
of our line shown in the graph is very similar to those
Tingey has published (Tingey 1979).

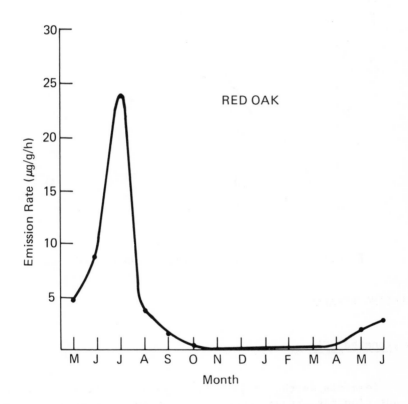

Figure 8.   Emission rate for red oak.

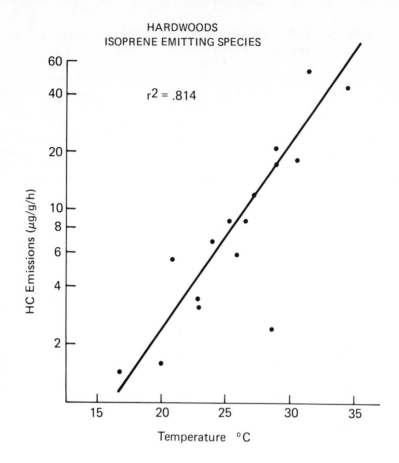

Figure 9.   Isoprene-emitting hardwood species.

CLOSING REMARKS

Oxygenated HC's

     Comments on the subject of oxygenated HC's are in order.
A question has arisen concerning whether oxygenated HC's are
emitted in appreciable quantities by natural vegetation.
Method verification studies have been performed for
oxygenated species similar to those described earlier for
pure HC's. The collection, concentration, and
chromatographic methods used for pure HC's are satisfactory
for oxygenated species containing an ether, aldehyde or
ketone linkage, but are unsatisfactory for alcohols. Acids
have not been studied. Table IV lists the oxygenated HC's

38

verified as present in vegetative emission samples.
1,8-Cineole and camphor were identified in sagebrush
emissions; methyl chavicol was released by pine. These types
of oxygenated HC's can be measured without problems.

TABLE IV. OXYGENATED TERPENES IDENTIFIED IN EMISSION SAMPLES

Compound	Vegetation Type
1,8-Cineole (ether)	sagebrush
Camphor (ketone)	sagebrush
Methyl chavicol (ether)	pine

Terpene Alcohols and Silylation

Tissue extracts from species sampled for HC emissions in
Pennsylvania in the summer of 1979 revealed the presence of
terpene alcohols (Table V). These compounds were not found
in emission sample volatiles, which suggests collection
and/or chromatography problems. The storage stability of
alcohols in stainless steel canisters is poor, and at low
concentrations (ppb), alcohol lifetime is less than 4 h.

TABLE V. OXYGENATED TERPENES IDENTIFIED IN TISSUE EXTRACTS

Compounds
2,4-Hexadadiene-1-ol
Benzyl alcohol
3-Hexene-1-ol
Linalool
Nerol
Benzaldehyde
Bornyl acetate
Methyl salicylate

Quite recently, efforts to circumvent this problem have begun, involving formation of a silyl derivative of the alcohol (which our initial laboratory studies suggest may be useful). Figure 10 shows a chromatogram obtained by injecting microgram quantities of a mixture of methanol, ethanol, propanol, butanol, and pentanol. Note that the peak's tail and the chromatography is not very good. As shown in Figure 11, using lower concentrations causes the first two peaks, methanol and ethanol, to be lost. Thus, obtaining quantitative data for alcohols will be difficult as lower and lower concentrations are sought.

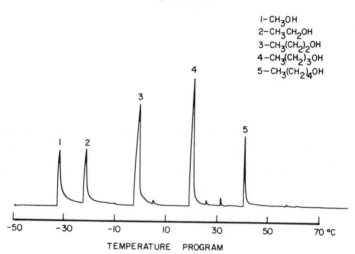

ALCOHOLS AT 4 μg/INJECTION

1—CH_3OH
2—CH_3CH_2OH
3—$CH_3(CH_2)_2OH$
4—$CH_3(CH_2)_3OH$
5—$CH_3(CH_2)_4OH$

TEMPERATURE PROGRAM

Figure 10. Chromatogram obtained by injecting five alcohols in sample.

Silyl derivatives of alcohols can be formed in the gas phase as shown in Figure 12. This chromatogram shows the silyl derivatives of the five alcohols shown in Figure 11. The derivatives were formed by combining the alcohols and a silylating agent in a canister in the presence of dry air. The peaks corresponding to the silyl derivatives are narrow and symmetrical, peaks #2 and 4 being the silyl derivatives of water. Peak #2 is monosilylated water; peak #4 is disilylated. Silylation appears to be a very sensitive technique for determining the presence of dry air.

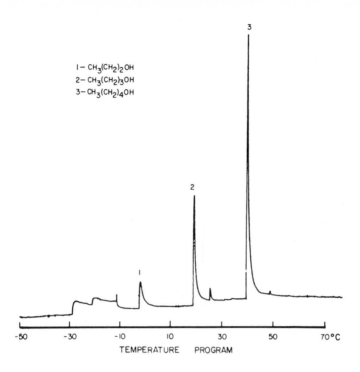

Figure 11. Chromatogram showing loss of methanol and ethanol
peaks at an injection concentration of 20 mg.

One of the main problems with the silylation procedure
is the large amount of water present in ambient atmospheric
samples. The chromatogram shown in Figure 13 is from a
canister containing silylating agent, roughly 50% relative
humidity and the five alcohols mentioned. Encouragingly, the
alcohols seem able to compete for the silylating agents;
because silylated alcohols are found. But certainly a large
part of the chromatogram is disrupted by silylated water
peaks. This problem might not be as serious for derivatized
alcohols with retention times much greater than the silylated
water derivatives.

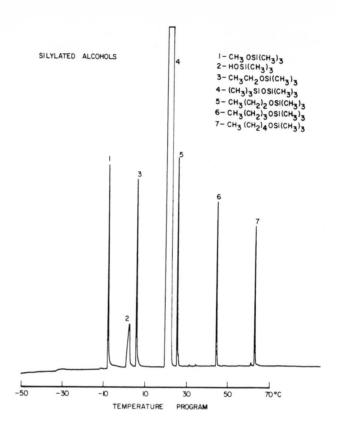

Figure 12.   Chromatogram showing silyl derivatives
of alcohol formed in the gas phase.

Two final points should be emphasized in closing. First
of all, ambient terpene concentrations are very low. Our
emissions studies have shown that the major components
emitted are actually terpenes, normally accounting for 80 to
85% of the HC's measured. The second point is that
oxygenated species appear to comprise a minor part of
vegetative emissions.

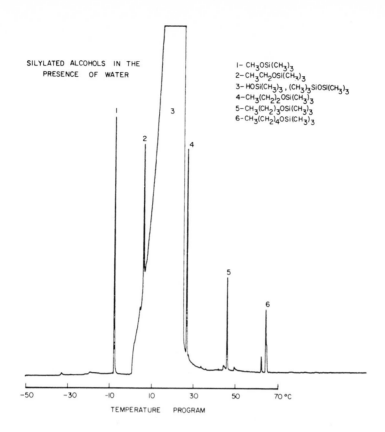

Figure 13.   Chromatogram showing silylated alcohols
in the presence of water.

REFERENCES

Tingey, D. T., M. Manning, L. C. Grothaus, and W. F. Burns.
1979. The influence of light and temperature on isoprene
emission rates from live oak. Plant Physiology. 47:112-118.

Went, F. W. 1960. Blue hazes in the atmosphere. Nature.
187:641-643.

Went, F. W. and R. A. Rasmussen. 1965. Volatile organic
material of plant origin in the atmosphere. Proc. of the
Natl. Acad. of Sci. 53:215-220.

Zimmerman, P. R. 1979. Testing of Hydrocarbon Emissions
from Vegetation, Leaf Litter and Aquatic Surfaces, and
Development of a Methodology for Compiling Biogenic Emissions

Inventories. EPA-450/4-79-004. U.S. Environmental
Protection Agency, Research Triangle Park, North Carolina.

DISCUSSION OF PRESENTATION

DIMITRIADES: Were the relative rates shown in Table I
determined in the presence of sunlight and NO_x?

WESTBERG: The relative rates shown were measured in a
laboratory chamber using lighting that simulated natural
sunlight with oxides of nitrogen present.

STEVENS: What parameter was measured in the kinetic studies?

WESTBERG: We measured the disappearance rate of the HC by
gas chromatography.

BUFALINI: How long was the loss of terpene followed in the
kinetic studies?

WESTBERG: Whenever possible the reaction was followed for at
least two half-lives. With the very fast reactions this
wasn't possible.

OLLISON: Is methyl chavicol shown in Figure 3?

WESTBERG: No.

RASMUSSEN: From what plant species was the chromatogram
obtained that's shown in Figure 3?

WESTBERG: This was an emission sample from Blue Spruce.

RASMUSSEN: Which peak in Figure 4 corresponds to benzene?

WESTBERG: It is one of the peaks about half way between
isoprene and toluene.

DIMITRIADES: Is this the system that you are using routinely
for analysis of urban air samples?

WESTBERG: Yes.

STEVENS: What volume of air do you pass through the trap
during the cryogenic concentration step?

WESTBERG: Anywhere from 100 to 1000 cc. I would say that we
generally use 500 cc.

STEVENS: How do you prevent water condensation with a sample size that large?

WESTBERG: At these volumes we have not had water problems. If you use more than one liter, trap freeze-up problems can be expected.   Water does not disrupt the chromatogram when SE-30 GC columns are used.

VOICE FROM AUDIENCE: Have you checked storage stability of the light HC's in stainless steel canisters?

WESTBERG: These were not included in Table II because of space limitations.   The C_2-C_5 HC's have been included in our stability studies and we find that they exhibit minimal loss over a 2-week period.

RASMUSSEN: In your storage stability results [Table II] you show a delta of 4%.   What would you say is the repeatability of your measurement?   Can you measure isoprene to within plus or minus 1%?

WESTBERG: We feel that our reproducibility is 10% for repeated analyses.

RASMUSSEN: So, for 2-methylpentane where you report a loss of 2%, there may be no loss at all?

WESTBERG: That's correct.

OLLISON: Is the initial concentration shown in Table II a calculated or measured value?

WESTBERG: It is an analytical value measured immediately after filling the canister.

DIMITRIADES AND STEVENS: The question is - is there any loss of the terpenes when they are added to a canister containing other reactive materials?

WESTBERG: This is a question I intend to address next [see text related to spiking of canisters with excess NO].

WALKER: What type of storage conditions were used in the canister stability studies?

WESTBERG: The canisters were kept at room temperature.

WALKER: Did you heat the canister before withdrawing a sample?

WESTBERG: No.

VOICE FROM AUDIENCE: Did you make 7-and 14-day measurements on the canister that had been spiked with NO to remove O_3?

WESTBERG: No, these measurements were made the same day only.

VOICE FROM AUDIENCE: How long did the canisters and Tenax samples in Table III stand before analysis?

WESTBERG: The samples were run the same day as collected.

VOICE FROM AUDIENCE: Was anything added to the canister or Tenax to quench O_3?

WESTBERG: No.

RASMUSSEN: How much air did you pass through the Tenax?

WESTBERG: We passed 200 cc through the Tenax and 200 cc from the canister.

SEILA: In the seasonal emission rate studies did you make measurements on the same tree each month?

WESTBERG: Yes, these data include one sample a month from the same tree.

LUDLUM: Are the emission rates shown in Figure 7 total terpene or only α-pinene?

WESTBERG: Total terpene.

HULL: Did the mix of terpenes change at all during the year?

WESTBERG: This is the next subject I will address [see text concerning seasonal changes in needle oil and oleoresin content].

OLLISON: How did the methyl chavicol vary with season?

WESTBERG: Methyl chavicol varied considerably. In most monthly samples, the chavicol peak was very small, but in the fall it comprised 30% of the total emissions.

RASMUSSEN: What happened in the fall of the year when the Ponderosa pine lost its needles, which were its 5th-year needles? Did you have any major flush in September or October - the oils have to go somewhere?

WESTBERG: As shown in Figure 7, there does appear to be a small increase in emissions in the fall, but it is small compared to the springtime increase.

ARNTS: In your survey of the literature, have you noticed any variation in terpene resin content for the same tree species growing in different geographical areas?

WESTBERG: Yes, there is variation with geographical location, but since we didn't make resin and needle oil analyses in conjunction with the volatile emission measurements, the exact source of the volatile emissions is rather uncertain.

RASMUSSEN: Were the data used to construct the temperature/ emission rate correlation shown in Figure 9 obtained by the bag enclosure method?

WESTBERG: Yes.

SEILA: Do the data shown in Figure 9 come from several different species?

WESTBERG: Yes.

RASMUSSEN: From what vegetation species were the oxygenated compounds shown in Table V obtained?

WESTBERG: Right off hand, I'm not sure of the exact compound/species relationships. We collected tissue extracts from about 30 vegetation types during this study. I have the information with me and will be glad to share it with you.

RASMUSSEN: I just wondered if you had oak in there, because the major pheromone that turns on the moths from oak, and all oak leaves do this, is the aldehyde 3-hexenal.

DIMITRIADES: It is my understanding that, by using your chromatographic system, you can detect terpenes other than isoprene and the pinenes if they are present in ambient air.

WESTBERG: This is correct.

DIMITRIADES: I would like to ask Bill Lonneman if he is using the same chromatographic system?

LONNEMAN: Our column system is different, but we can detect the other terpenes.

DIMITRIADES: You can see them, too. So, if terpenes had been present in the Houston atmosphere during the 1978 field study there you would have seen them.

LONNEMAN: Neither the Washington State nor EPA groups found terpenes in the ambient Houston atmosphere.

DIMITRIADES: I guess this answers the question that Rasmussen raised earlier - whether or not terpenes represent only 50% of what may be emitted into the atmosphere by vegetation.

RASMUSSEN: The only point I would make on the Ponderosa foliage on Moscow Mountain, versus the pine foilage, and the slash pine and the loblolly, and the pinus strobus, the white pine of the East Coast, is we've always found that the pines on the East Coast, associated with a mesenthytic-type forest, are prone to release more material.

We see more material coming out of these, and this is what Zimmerman's study has shown. There's something very stingy about the emissions coming out of the Ponderosa pine in the far West. They are just more adapted to a xeric environment. There are more sclerenchyma cells in their tissue, they are much higher and their resin ducts are very much deeper into the tissue. And they're just not as prone to release terpenes as some of the other pines.

But it still raises the question, such as pertaining to the foilage analysis of Zimmerman's in Tampa, Florida, of why does the rather high loading of paraffinic and aromatic material occur. His samples weren't 80-85% isoprene and monoterpenes, but only on the order of something less than 50%.

ZIMMERMAN: The major point is that you're talking about a tree emission. And when we talk about the Tampa inventory, we're talking about soils, leaf litter, and surface waters. Even though those emissions are quite small, when you average the percent coverage of the total area and you multiply times the square feet of ground surface, you can get quite a bit of stuff that isn't terpenes.

So, when you're just talking about trees--and especially specific trees--you can identify most of them as olefins. And you can certainly see from the distribution maps that I showed yesterday, where we had an oak/hickory type forest, we saw mostly olefins. But where you have mixed vegetation and shrubs, it is difficult to identify anything. But for some, we identified virtually everything. So, it depends on the vegetation base.

SIEVERS: First, I want to say that I think, clearly, what we desperately need is more attention to compound-by-compound identifications. And you've very beautifully shown us a way to do that.

I would also point out, that with the advent of the new fused silica capillary columns, others have observed--and we have confirmed in very preliminary experiments--that many of these compounds that you normally would have to methylate, like phenols and carboxylic acids, can be done directly without derivativization in the fused silica columns. So, I think that gives us the possibility for learning a great deal more about maybe what's happening to the terpenes that are clearly being emitted but disappear very rapidly, and do not show up as any appreciable concentration in polluted cities.

The other observation that I have is that the work going on in Europe, and the European Research Center under Versino and his coworkers, has indicated that contrary to your perhaps general observation, O_3 doesn't build up in all cases when terpenes are photooxidized. I think you made the observation, that you expected emissions from the trees (terpenes) to behave similarly to the olefinic compounds emitted in automotive exhaust. I think some of the things that Versino and the others are doing would not support that contention. They take bags and irradiate them in the presence of little NO_x. In the case of emissions from Ponderosa pine, which were being studied when I was there, the increase in O_3 was not substantially above that of the blank. In the case of automotive exhaust, there was a pronounced difference. And these were comparable concentration levels of total olefins. So it's not, I think, fair to say that olefins ought to contribute universally in the same way to the oxidant problem. Further, in the case of individual terpenes that they studied, myrcene for example, there was a net decrease rather than increase in O_3 concentration at certain stages of their experiment. So, I just wanted to offer that observation that these experiments are at variance with the general supposition that you were making.

BUFALINI: I know that Hal could answer, but we'll have to move on. I will make a very quick comment on that. Hal has studied HC-to-NO_x ratios--O_3 as a function of HC-NO_x ratios--and indeed, the amount of O_3 that's being produced depends on that ratio.

As you suggested, the one figure he picked was a generality. But he is, and we are, well aware of the fact that production of a significant concentration of O_3 depends upon how much NO_x you have relative to the HC's.

12. RURAL NONMETHANE HYDROCARBON CONCENTRATION AND COMPOSITION

Martin A. Ferman. Environmental Sciences Department, General Motors Research Laboratories, Warren, Michigan 48090

ABSTRACT

General Motors has performed ambient air quality studies in rural areas over the past decade to monitor ozone levels and determine possible sources of ozone. Nonmethane hydrocarbon levels measured in these studies have typically been found to be 0.1 ppmC or less (as opposed to urban levels over 1 ppmC). A mobile laboratory housed the gas chromatograph and flame ionization detectors used to measure the hydrocarbon species, and was parked outside the forest canopy in an open field for each site monitored. Bag samples were also taken for conducting irradiation experiments. The results of these studies indicate that light olefins are depleted in rural settings, but a major exception was the high level of isoprene measured in Keysville, Virginia. Further studies are needed to distinguish biogenic from anthropogenic hydrocarbons, though the major anthropogenic hydrocarbons involved in urban photochemistry do not appear to be important in rural areas.

INTRODUCTION

General Motors (GM) has not been directly involved in making any biogenic emissions measurements; however, over the last ten years, ambient air quality studies have been conducted at between 25 and 30 different sites around the country. For all but one of the sites presented in this paper, our main interest was the study of ambient ozone levels and identification of possible sources of ozone. In this context, meteorological parameters and ozone precursors were also measured, including individual hydrocarbons (HC's) in rural areas.

Figure 1 shows the sites that will be discussed. The first rural studies were conducted in 1975 and 1976 at sites in south-central Virginia, near Keysville (sites 1 and 3 on the map); in extreme western Maryland at McHenry (site 2); near McKee, Kentucky (site 4); and finally, in the Pisgah National Forest in western North Carolina (site 5), east of Asheville. Site 2 was in the mountains, a little over 100 km southeast of Pittsburgh. The McKee site, in central Kentucky, was on the other side of the Appalachians from the two Virginia sites.

In 1978, a more remote location in central South Dakota (site 6), west of Ft. Pierre, was chosen. The most recent study, conducted last summer in southern Louisiana (site 7) will not be discussed due to incomplete validation of the data.

To compare data collected at these rural sites with data collected in several urban areas, this paper will also present results of studies conducted in Houston, Texas, and Denver, Colorado.

Denver was the only study not conducted primarily to assess ozone levels; it was a wintertime effort for studying urban aerosol and haze problems. All other studies were conducted during the summer photochemical season.

INSTRUMENTATION

GM Atmospheric Research Laboratory

Figure 2 shows the mobile laboratory on location near Busick, North Carolina. In addition to the HC measurements at this and all sites, we also measured ozone, nitrogen oxides, sulfur dioxide, fluorocarbons, visibility, and a complement of meteorological parameters that included: wind speed and direction, temperature, dew point, barometric pressure and others.

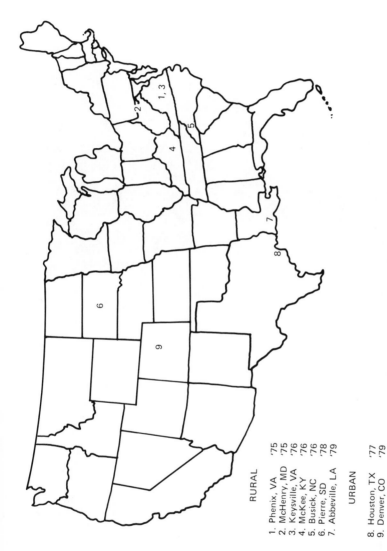

RURAL

1. Phenix, VA   '75
2. McHenry, MD   '75
3. Keysville, VA   '76
4. McKee, KY   '76
5. Busick, NC   '76
6. Pierre, SD   '78
7. Abbeville, LA   '79

URBAN

8. Houston, TX   '77
9. Denver, CO   '79

Figure 1.   Recent ARL field studies.

Figure 2 also shows the equipment for the irradiation experiments we conducted at several sites using used bags constructed from 2-mil Teflon. Ultraviolet (UV) transmission is very good through 2-mil FEP-Teflon. These experiments were performed in an attempt to quantify the potential of the ambient air to photochemically generate ozone. Unfortunately, problems with contamination greatly limited the usefulness of the bags in rural areas.

Figure 2. The Atmospheric Research Laboratory near Busick, North Carolina, September, 1976.

HC Analyses

Sampling--

All HC measurements were made in real time, aboard the mobile laboratory, using gas chromatography (GC) with flame ionization detectors (FID's). Ambient air was sampled at 10 m above the ground through the large glass sample tube. Figure 3 shows a diagram of the sampling system.

A 1-m, 6.3-mm diameter Teflon line extends from the center of the ARL's 10-cm sample duct to a metal bellows pump with a sintered metal filter on the intake. From the pump, the sample is put through a stainless steel and Teflon trap

in an ice water bath to lower the dew point and prevent
liquid water from accumulating anywhere else in the system.
Hydrocarbon-free air (from an Aadco generator) or calibration
(cylinder) gas may be introduced through an electrically-
operated switching solenoid mounted near the pump exit, so
that all calibration gases also pass through the water trap
and the rest of the sampling line. All valves, lines, and
fittings are constructed from either stainless steel or
Teflon. A back-pressure regulator mounted on the GC
maintains 8 psig of sample by venting excess immediately
downstream of the "cross" fitting that splits off sample to
the gas-sampling valves mounted inside the GC. Typical
sample flow from the pump is about 500 ml/min - most of which
is vented.

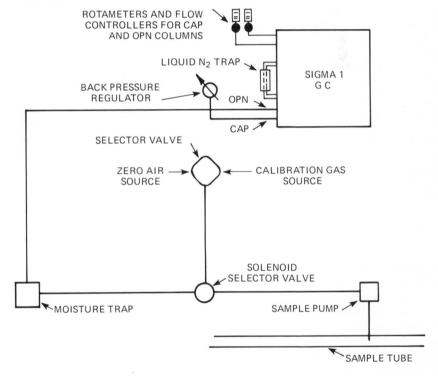

Figure 3. Schematic of ARL hydrocarbon analytical system.

The "cross" before the back-pressure regulator splits
the sample stream into thirds: 60 ml/min of air to both
gas-sampling valves, and the remainder through the regulator
to be vented. After this fitting, all lines, fittings,
valves, and loops are constructed from stainless steel. (The
only exception is the glass and Carbowax packing material in
the traps.) Regardless of the valve position (shown in
Figure 4), the sample air emerges from the valve and is

plumbed through a flow controller to a rotameter and vented. The normal, deactivated position of the valves allows the sample air to pass through a restriction directly to the flow controllers and rotameters. For 3.5 min, while the traps are cold, both valves are actuated and sample air is metered through the two traps, freezing out all organic compounds. Good control of the flow rates and the timing is essential since they determine the sample size (e.g., 60 ml/min for 3.5 minutes gives a 210-ml sample through each trap). Either parameter may be varied to change the sample size, although in practice variation is rarely necessary and is done only when the need for higher sensitivity requires a larger sample. Then the trapping time may be doubled or the flows increased up to 90 ml/min so that 450 ml of air flows through each trap. Trapping over 500 ml can present problems of ice plugging the traps.

Freeze-out Traps--

The trap assembly (Figure 5) was designed to be small, with minimum dead volume, so that it could be easily and accurately cooled and heated, reliably automated, and easily cleaned. Two sample traps, constructed from 20-cm x 3-mm thin-wall stainless steel tubing, were used to collect light HC's (C_2-C_5), and heavy HC's ($>C_4$), respectively. (The two samples are kept completely separate from the time they leave the "cross" to entering one of the valves.) The light HC trap is packed with Carbowax 20M 60/80 mesh, and the other trap is packed with 60/80 mesh glass beads. The traps are mounted in parallel about 2 cm apart through the center of the Teflon housing. They are electrically connected at one end, insulated from the GC and Teflon housing so that they can be wired in series for resistance heating. An iron/constantan thermocouple is silver-soldered to the center of the Carbowax trap and is connected to two temperature controllers - one for cooling and one for heating.

This system eliminates many problems associated with sample storage or collection on solid sorbents. Based on a variety of studies, all hydrocarbons are trapped at essentially 100% efficiency, with no detectable holdup on the trap on desorption. Analysis occurs within minutes of sample collection, and the whole sampling line is easily cleaned and kept clean, because no connections are ever made or broken. A routine check for contamination can easily be made by flipping a switch to sample clean air through the whole system.

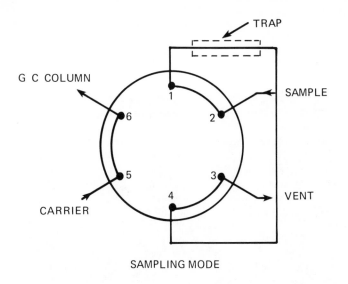

SAMPLING MODE

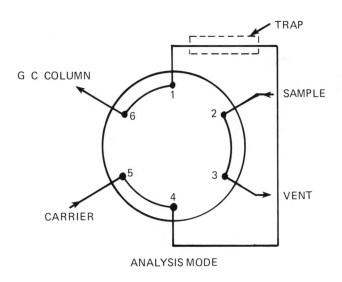

ANALYSIS MODE

Figure 4. GC valve positions.

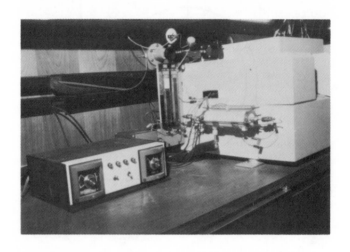

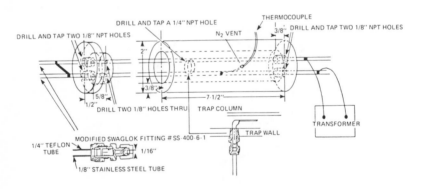

Figure 5. ARL cryogenic hydrocarbon trap.

Automation and Timing--

The GC is wired and programmed to start a run automatically when all temperatures and flows are correct. The events listed below are thus initiated:

Time (min)	Event
0.0	Cool trap, start sample flow
2.5	Start Trapping
6.0	Inject
6.1	Liquid nitrogen off
6.2	Trap heater on
7.0	Start GC temperature program
10.0	Trap heater off

First, the cold temperature controller is activated, which controls the liquid nitrogen solenoid to cool the trap and maintain the $-150°C$ trapping temperature. At the same time, the sample flow begins flushing out the sample lines. Power to the sample pump and the calibration solenoid is wired through a double-pull, double-throw switch so that only one device operates at any one time.

The trap is cooled down in two minutes by setting liquid nitrogen contained in Dewar flasks to vent at 22 psig, and by adjusting a variable restriction in front of the liquid nitrogen solenoid. At 2.5 minutes, both sample valves are actuated, directing sample flow through the cold traps. At 6 min, the valves are deactivated, injecting a small dead volume of air. The temperature control is switched to the heating controller 20 sec later, which powers a transformer providing 3 VAC across the traps, heating them to $100°C$ in about 10 sec. The traps are kept at $100°C$ for 3 min. (Only the steel traps are heated; the housing slowly warms to room temperature.)

Analysis/Chromatography--

Each sample valve is plumbed directly to an analytical column, with conditions listed in Table I. Analysis of C_2-C_5 compounds is performed on a 6-m x 1-mm stainless steel column packed with Durapak OPN 80-100 mesh Porasil C. Heavy HC's are separated on a 50-m x 0.25-mm ID glass capillary column coated with OV-101 methyl silicone (purchased from Perkin-Elmer). Of course, other columns may be used to solve specific analytical problems, but these two are the best that

59

GM has found for routine ambient analyses (when only one GC is available for HC's).

TABLE I. GAS CHROMATOGRAPHIC CONDITIONS

Feature	Description	
Columns	Durapak OPN/Porasil C 6 m x 1 mm stainless steel	OV–101 WCOT 50 m 0.25 mm glass
Helium	25 ml/min	3 ml/min
Temperature program	Initial hold at −70°C 4°/min to 100°C 8°/min to 120°C	
Detection	Flame ionization (H_2, air)	
Minimum detectable conc	0.1 ppb C	
Sample size	200 ml	

Error Analysis

Accuracy	± 10%	
Precision (concentration)	± 1% (1 σ)	
	± 0.1% hexane ± 3.0% acetylene	
Reproducibility of retention times	± < 0.3%	10 runs over 1 week 1 σ ≈ 4 sec

High-purity helium is the carrier for both columns. The gas passes through two pressure regulators (the second at 90 psig), and then through two molecular sieve traps. Flow rates are 25 ml/min through the packed column, 3 ml/min through the glass capillary, and 30 ml/min make-up for the

capillary. The oven temperature is programmed to hold at
-70°C until 1 min after the traps are thermally desorbed, and
then to increase at 4°C/min to 100°C and 8°/min to 120°C
where it is kept for 10 min before cooling down for the next
run.

Two flame-ionization detectors are used with air and
hydrogen flows adjusted for the optimum signal/noise ratio.
The minimum detectable concentration, corresponding to a peak
height twice the noise level, is about 0.1 ppbC and varies
with different compounds. However, interference sometimes
occurs from a small number of peaks this size or slightly
larger in blank runs (with no injection), presumably coming
from either the carrier gas, trap, or other material in the
system.

Data Processing--

The Sigma-1 GC microprocessor integrates the peak areas,
identifies the peaks and prints out a complete report
containing calibrated (ppbC) values for all peaks. Based on
studies with the dilution system that will be described
later, peak area is linearly related to concentration and
correlates well with peak height over the concentration range
of interest.

Peak identifications are based on retention times, which
are very reproducible. For ten analyses of the calibration
cylinder run over a 1-week period in the latest field study,
retention-time reproducibility was better than $\pm$ 0.3% (1σ)
for all light HC's. Identifications are checked against
relative retention times, usually based on dominant peaks
expected at a particular site (such as propane, butane,
isopentane, and/or toluene). Typically, chromatographs of
ambient air contain about 100 well-resolved peaks greater
than the 0.1 ppbC minimum detectable concentration. We have
identified about half of these, corresponding to
approximately 75% of the total mass.

The raw data from the Sigma-1 printouts are read into
the GMR IBM-370 computer for further calibration and
analyses. Each concentration is adjusted according to
individual response factors and the results of frequent
analyses of a calibration cylinder containing over ten
different compounds, from ethane to m-xylene. Individual
response factors are determined empirically for about twelve
different compounds and estimated for the rest.

Calibration and Quality Control--

Repeated analyses of a calibration cylinder containing
ten compounds at 1 ppmC each indicated an average
reproducibility of $\pm$ 1% (see Table I). This value varied for
different compounds, from $\pm$ 0.1% for the analyses of hexane
to $\pm$ 3% for acetylene.

When collecting ambient data in the field, the
calibration cylinder is run once a day or once every other
day. If results differ by more than 5% from the previous
calibration run, another analysis is made. During a recent
study, 24 analyses of this cylinder were performed over a 6
week period. After normalizing on propane to account for
day-to-day changes in ambient pressure and temperature, as
well as for different air and hydrogen gases, the standard
deviation averaged $\pm$ 2% for ethane, ethylene, acetylene, and
isopentane (the light compounds in the cylinder).

All of our calibration cylinders are purchased with a
stated accuracy of $\pm$ 2% from the Scott Laboratories. Every
six months we reanalyze every cylinder using one of several
methods, most of which involve dilution of pure compounds.
Liquid HC's are usually checked by injection with a
microliter syringe into a 8-m^3 stainless steel chamber. The
best method we have found for gases (with vapor pressures
above 780 mm at room temperature) employs a two-stage
dilution with zero air, using mass flow controllers. A
diagram of the dilution system is presented in Figure 6.

A number of advantages exist with this dynamic dilution
method. Several gases can be blended at the same time to
make practically any concentration down to 0.01 ppm with a
continuous output up to 20 liter/min, which can be maintained
for days. All variables that influence the concentration
(flow rates, temperature, pressure) can be accurately
(usually $< \pm$ 5%) and conveniently determined using
conventional techniques. Once the concentrations are set up
and calibrated, they can be changed in a matter of seconds,
simply by changing the set point on a mass flow controller.
This system allows multi-point calibrations of each gas to
check linearity and response factors, in addition to
calibrating our reference cylinders.

Because so many variables are involved in this method,
the absolute accuracy depends on the care with which each
parameter is checked. Typically, each of the four flow
measurements can be made within 2%, and the precision of the
GC analyses also is $\pm$ 2%. The purity of the supply gases, at
flow rates that produce ppm concentrations, adds less than
0.5% uncertainty. Absolute accuracy is estimated at $\pm$ 10%.

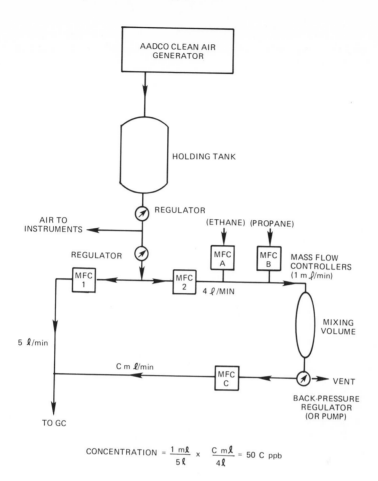

CONCENTRATION $= \dfrac{1\ m\ell}{5\ell} \times \dfrac{C\ m\ell}{4\ell} = 50\ C\ ppb$

Figure 6. Flow dilution calibrator.

Examples of Ambient Analyses--

Figure 7 shows a chromatogram from the packed column of light HC's in rural Louisiana. This site was neither pristine nor remote, as can be seen by the relatively large paraffin peaks (presumably leaks from natural gas wells). This chromatogram illustrates our good sensitivity, resolution, and baseline noise and stability.

A sample chromatogram of HC's from the capillary column is shown in Figure 8. This sample was actually one of the cleaner samples from Denver. Although the peaks are a little lower than average for Denver, this chromatogram is typical with over 100 well-resolved peaks. Most ambient samples are as complex, and peak identifications based on retention times can be very difficult.

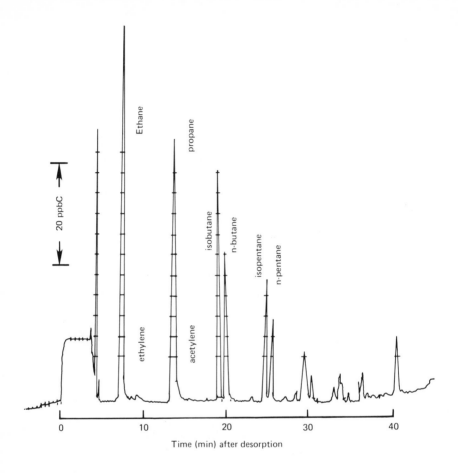

Figure 7. Chromatogram of ambient air in Abbeville, Louisiana, August 1979.

As mentioned, retention times are very reproducible with the microprocessor control (+ 0.3%), averaging about four seconds for toluene. However, several compounds can elute during this time within this region of the chromatogram. Although not indicated in Figure 8, most pinenes elute near nonane, while isoprene is near n-pentane.

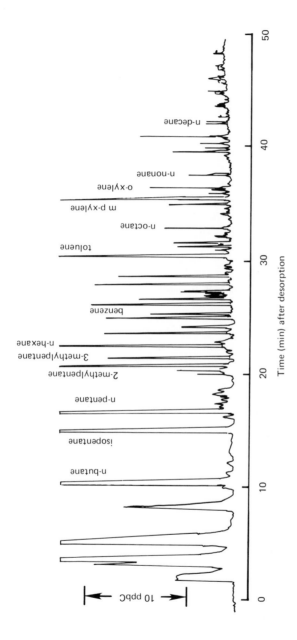

Figure 8. Chromatogram of ambient air in Denver, Colorado, December 1978.

65

For a well-resolved peak such as α- or β-pinene, the minimum detectable concentration for this sample size is approximately 5 ppt compound, corresponding to twice the baseline noise. Again, since the primary interest in conducting these studies was not to quantify biogenic hydrocarbons, retrospective recovery of chromatographic data on these biogenic compounds has been quite difficult. The problem is not one of sensitivity, but of identification.

RESULTS AND DISCUSSION

Table II lists the average concentrations of light HC's that have been positively identified for urban, suburban, and rural areas. Houston, of course is noted for its petrochemical industry and associated HC emissions. Our GM location during the 1977 HAOS program was across town from the refineries, in a suburban setting. The levels of most pollutants at this site were about half as high as the levels that have been measured in most urban areas (such as the Denver sites). At the rural Keysville, VA site, we measured HC levels one to two orders of magnitude lower than typical urban levels. Except for one case that will be mentioned later, the concentrations measured at Keysville are similar to those measured at other rural sites and by other researchers.

TABLE II. INDIVIDUAL HYDROCARBON LEVELS (ppbC)

	Urban (Denver)	Suburban (Houston)	Rural (Keysville)
Methane	2100	2020	1650.0
Ethane	75	22	2.9
Ethylene	54	14	0.3
Acetylene	50	14	0.9
Propane	101	43	2.8
Isobutane	58	32	0.9
N-butane	150	72	2.1

The South Dakota site was our most remote site and all pollutant levels there were lower than those at the rural sites in the East. Hydrocarbon levels at the Pierre, South Dakota site were so low that we had serious problems with small amounts of contamination in the carrier gas. Though complete statistics were not compiled for Pierre HC's, no peak exceeding 4 ppbC was noted, and, on the average, levels were about half of those reported for Keysville.

The few rural heavy HC's that were identified and quantified include benzene, toluene, xylenes, and terpenes, most of which were detected in the 0.5 to 2 ppbC range at the rural sites in the East. In urban areas, levels of aromatics were typically 30 to 80 ppbC. Terpene peaks could not be positively identified in urban chromatograms.

In general, for rural sites in the East, the total nonmethane HC was approximately 100 ppbC. Of this amount, about 10 to 20% was light paraffins, 5 to 10% terpenes, and about 5% was aromatics. Contrasting measurements were taken at the Louisiana site in the summer of 1979, where light paraffins at times exceeded 1 ppmC. These extremely high concentrations were probably due to upwind gas wells. The only other large exceptions to the above general results were the high concentrations of isoprene consistently measured at the Keysville site – averaging 13 ppbC, with the maximum over 150 ppbC. Isoprene alone accounted for over 10% of the nonmethane HC; no other compound was detected at levels even approaching those for isoprene (which we carefully validated). At other rural sites, isoprene levels were generally 1 or 2 ppbC.

Though the Keysville site was surrounded by hardwood forests, sampling was not performed in the canopy there or at any other site. The forest contained a high percentage of oak trees and the weather was frequently hot and sunny, but the levels of isoprene were still surprisingly high. (However, over the 50-m distance between the woods and the sampling station, isoprene should transport better than terpenes, which react faster with ozone).

One unique factor at the Keysville site was tobacco. The laboratory was set up near an open field that was to remain fallow all summer. A few weeks later, however, the resident farmer planted tobacco all around the site. Nevertheless, tobacco does not appear to be a particularly strong isoprene emitter.

The next figure (Figure 9) shows the diurnal distribution of isoprene levels at Keysville and McKee. Although isoprene concentrations were generally higher in the late afternoon, too few data points may have been collected for a representative curve, (i.e., some hours may have been dominated by one or two extreme values).

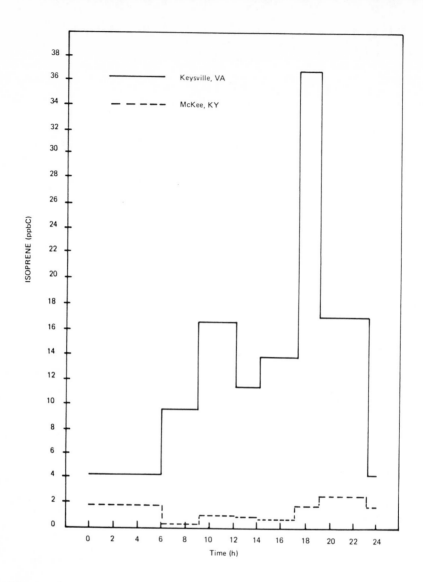

Figure 9. Diurnal plot of ambient isoprene concentrations.

In Table III, the same three sites are again compared. The first item on the table illustrates the large differences in total nonmethane HC concentrations at the three sites. In addition to differences in the total, significant differences also occurred in composition. One obvious difference is the lower relative amounts of light olefins in rural and urban areas, with intermediate compositions at the surburban site. This apparent loss of light olefins was observed for all C_2-C_4 olefins at all rural sites; in fact, C_3 and C_4 olefins

68

were rarely detected in rural areas. Their high reactivity (as compared to paraffins) would be expected to produce shorter lifetimes. As for isoprene, the previous discussion reveals that this C_5 diolefin was definitely not depleted, at least at one rural site.

TABLE III. COMPARISON OF URBAN/SUBURBAN/RURAL LEVELS

	Urban	Suburban	Rural
Large Dilution NMHC (ppm C)	1.2	0.6	0.1
Loss of Olefins			
Ethylene/acetylene	1.1	1.0	0.3
Ethylene/ethane	0.7	0.6	0.1
Little loss in complexity (No. of peaks limited by resolution)			

Even in rural areas where HC concentrations are one to two orders of magnitude lower than urban levels, the chromatograms may still contain 100 well-resolved peaks. For our analyses - limited to only one GC with two columns - positive identification of any compounds eluting past isoprene was very difficult. Consequently, quantification of total natural or anthropogenic compounds was not possible. (Hence the claim by others that the one or two compounds measured represent all biogenic compounds is questioned.)

HC Reactivity

Although all the HC's in ambient air cannot be identified or specified as anthropogenic or biogenic, typical levels of many species in both urban and rural air are known. A model would be desirable that could specify just how the compounds react, and what products are formed. A major issue is the question: how significant are biogenic emissions in terms of the net production of ozone and visibility-reducing aerosol? Accurate quantitative models may still be far in the future.

To put the presented ambient measurements into perspective, a first approximation of HC reactivity (that is easily available) would be to compare the rates of reaction of individual HC's with OH. This reaction is important as a chain-initiating step in the mechanism that results in the production of ozone and oxygenated organics. For many HC's, reaction with OH determines their lifetime in the atmosphere, although for some of the more reactive species, like the terpenes, the reaction with ozone is probably more important.

Table IV lists typical levels for several compounds and groups of compounds and compares these species based on their reactivity with OH. The last two columns list the product of the concentration and the rate constant. These values are not meant to represent ozone-making potential directly, since this reaction is only the first step in the mechanism. Larger molecules may produce more radicals that can eventually oxidize more NO to NO_2 to produce more ozone.

TABLE IV. HYDROCARBON REACTIVITY WITH OH

Species	Typical Concentration (ppb C)		Conc x k(OH) (sec^{-1})	
	Rural	Urban	Rural	Urban
Methane	1650	2100	0.3	0.4
Carbon monoxide	150	2480	1.3	21.0
Alkanes C_2-C_4	9	392	0.1	3.8
Alkenes C_2-C_4	1	73	0.1	4.7
Isoprene	1-10		0.4-4.0	
Terpenes	≈ 5		1.1	
Aromatics C_6-C_8	≈ 2	156	0.1	5.3
Formaldehyde	≈ 4	≈ 10	1.0	2.5

Both CO and CH_4 are included in Table IV for comparison purposes because of their large background levels (compared to other pollutants). Methane, easily the most prevalent HC, is also the least reactive, and CH_4 oxidation is generally considered negligible. Carbon monoxide appears to be the most significant compound in terms of OH removal, which is not surprising. Several recent papers claim that CO levels determine the OH background concentration in the troposphere.

The high levels of ozone seen in urban areas during the summer are due almost entirely to photochemical reactions initiated by the oxidation of reactive HC's (CH_4 oxidation is unimportant). In rural areas, all light HC's, whether saturated or unsaturated, appear even less important than methane. Rural HC's that could be significant in this scheme are isoprene and terpenes. The concentration range listed for isoprene is the approximate range observed between average levels at all eastern sites.

The formaldehyde concentrations listed in Table IV were based on a limited number of measurements made in 1976 using the chromotropic acid/bubbler technique. Measurement of any other oxygenated compounds was not attempted.

SUMMARY

Several main results of our measurements of ambient nonmethane HC's in rural areas of the U.S. should be reiterated. First, the levels are low - typically 0.1 ppmC or less, as compared to urban levels over 1 ppmC. Furthermore, HC composition in rural areas appeared depleted in light olefins. The percent composition of ethylene and propylene was lower in rural samples than in urban, and rural levels of butenes were frequently below the limits of detection.

One major exception to these general findings was the high level of isoprene measured at Keysville (where one sample contained 150 ppbC of isoprene).

For a first approximation based on OH reactivities, the major anthropogenic HC's involved in urban photochemistry do not appear to be important in the local photochemistry of rural areas. Rather, biogenic terpenes and isoprene seem to have the largest potential for the generation of ozone or aerosol.

All of our measurements were made in realtime, at sites carefully chosen as remote from any major pollutant source. None of the sites was located directly in a forest canopy (where the levels of biogenic pollutants may be considerably higher than they are in open fields). Because of the complexity of the ambient data for heavy HC's (compounded by the limitations of only one instrument), we were unable to identify a sufficient number of compounds to allow quantitative separation of biogenic pollutants from anthropogenic ones. Identification of every significant

71

species is needed, including oxygenated compounds. Future
studies include using a mass spectrometer and additional
columns, adsorbents, and detectors.

DISCUSSION OF PRESENTATION

BUFALINI: I'd like to make a comment about your Keysville
site. As you know, there seems to be some extremely high
concentrations of isoprene. And at least, from what we've
seen for the IBP site, whenever you get away from the canopy,
the hydrocarbon concentration goes down exponentially - I'm
talking about the α-pinene concentration. I can't but help
wonder that if you were to go over your data, you might see
some wind dependence.

Your Keysville number is extremely high. We haven't
seen one that high before. If you can't relate it to wind
direction, then perhaps some wood was being cut in the area,
or something like that.

FERMAN: Yes, I think you're right. Everyone is reporting a
lower concentration of terpenes outside the canopy - that we
did see. Even at the time we saw a lot of isoprene, the
terpene concentrations were still in the 1 or 2 ppbC range.

The wind dependence would be tricky, because even though
the woods were closer on one side, on the other side was a
small field and more woods. We did see a meteorological
dependence - what appeared to be higher isoprene on the hot,
sunny afternoons (particularly the hot afternoons).

OLLISON: You might have the field radiating, and the thermal
coming up in the field and pulling out air from underneath
the cool canopy - what comes up.

BUFALINI: I think that it won't work out.

OLLISON: Well, they say in San Francisco that it comes down
the slopes. Same sort of thing.

ROMANOVSKY: Did you see any significant photochemical
activity in your IR radiation test?

FERMAN: That is something I intended to mention. We've done
a lot of these studies in urban and rural areas, as well as
in the lab. The first ambient irradiations we attempted were
in 1976, at the rural sites. And we've been doing it every
year since then, and we've learned a lot. Nelson Kelly has

conducted most of the studies for the past two years and can give you more details.

So far we haven't been able to get any significant results on the photochemistry in these rural areas, where contamination from the bags and other problems are at least as important as the ambient pollutants. We have had some good results from the irradiation studies conducted in urban areas, and with urban mixes.

Like I said, Nelson can give you a lot more information on that. But I'd be very cautious—

ROMANOVSKY: I'm concerned about the rural measurements.

FERMAN: We found a lot of problems with contamination from the bags at levels that are insignificant in a city, but are very significant in rural areas. Also, you have to be very careful, even in a city, how you clean up the bags.

ROMANOVSKY: Including level of contamination, what was the level of ozone?

FERMAN: Well, in the rural area, the levels we saw in the bags were a little bit higher, as I recall, than in the ambient air. Typically, on sunny days, we could generate a tenth of a part per million of ozone in the bags; while we were seeing 80 ppb outside. Again, that's in the eastern part of the country.

Ozone levels in South Dakota, both ambient and in the bags, were even lower. Ambient levels at Pierre were around 50 parts per billion.

KELLY: Excuse me, but the point is that you didn't have to put in anything; you could put zero air in the bag with a little bit of NO_x and make a lot of ozone.

RASMUSSEN: But the question really is, how long does that go on in the bag? I mean, how do you precondition your bags? If I can just preface this, that we have irradiated bags for as much as 180 days with continuous irradiation, 24 hours a day. You can select certain of these Teflon bags, and after a few days of irradiation, they do reach a zero point of releasing any further hydrocarbons into the interior of the bag, or producing any ozone. Even more significantly, the best crutch of checking on the quality of the bag is to see how much CO the bag is generating.

FERMAN: That's a very important point.

RASMUSSEN: And you really can get some of the bags down, but you have to put quite a bit of work into preselecting and preconditioning the bag.

KELLY: What we found is that new Teflon material – what we thought would be the cleanest (never having seen NO or anything) – does release CO. And it must be from the Teflon. It's not due to any photochemical reaction in the bag that degrades hydrocarbons to CO. The new material, for the first couple of days, releases large amounts of CO. It must be a photochemical reaction in the Teflon material.

RASMUSSEN: Well, that's really not quite the case, because you can clean up a bag, get zero CO, flush it with ambient air or suburban air, and start the CO process all over again, as you absorb on the wall.

KELLY: No, I'm talking about new material. For the first couple of days, CO production does go down.

BUFALINI: I think you can get two schools of thought, here, and you won't resolve the issue at all.

KELLY: I think we're in general agreement, though.

BUFALINI: As a matter of fact, if you do clean up the Teflon, over prolonged periods of irradiation, what Kelly says is probably right. But if you let the bag sit for two or three weeks, and then go back to it, you have to go through the whole acclimation process again. And it's due to the fact that it's coming off the Teflon, rather than permeation into the bag.

RASMUSSEN: Well, I think I agree with that.

BUFALINI: And we have scads and scads of evidence.

RASMUSSEN: Well, there are some real rotten grades of Teflon on the market, or Tedlar, but the point that I'm concerned about is that after you've used the bag once, you've got to go through this cleanup procedure again before further use.

BUFALINI: The other problem you would run into is whether or not it's a 2-mil or 5-mil bag.

It turns out, a 2-mil cleans up much better than the 5-mil. However, the 2-mil is so much more fragile, that once you develop a hole in it, the bag isn't any good anyway.

VOICE FROM AUDIENCE: Yeah; we've troubles with ripping.

BUFALINI: One more question, since we're running out of time.

KAISERMAN: You mentioned that you did some aerosol sampling at the sites. Would you have any information on that?

FERMAN: At the rural sites, for the most part, detailed analyses of the aerosol were omitted. In the earlier sites, we ran high-volume samplers for TSP and some radionuclides, like beryllium-7, which we were interested in as an indicator of stratospheric ozone transport.

For the more recent studies, we also have sulfate and nitrate data from low-volume samplers, with 4-hour time resolution. And we have a very extensive aerosol data base from the study in Denver. That study was primarily aimed at aerosols, and I can tell you more about it later, if you're interested.

13. HYDROCARBON SAMPLES FROM TEMPERATE GLACIERS
OF THE NORTH CASCADE MOUNTAINS

H. William Wilson. Department of Chemistry, Western
Washington University, Bellingham, Washington 98225

ABSTRACT

Hydrocarbons from temperate glaciers in the North
Cascades of Washington State have been analyzed using several
analytical techniques. The 50 to 60 compounds detected are
determined to be complex mixtures of partially-oxidized
hydrocarbons that resemble organic fractions of aerosols,
petroleum or coal bitumens, or soils and sediments. The
presence of certain polycyclic aromatic hydrocarbons suggests
these organics originate from precipitation fallout of both
anthropogenic and biogenic combustion products. Selected,
isolated treeline vegetation in the North Cascades is being
monitored to further investigate biogenic hydrocarbons
emitted from early spring to late fall.

INTRODUCTION

During the past 15 months, our laboratory has been
recovering and analyzing hydrocarbons (HC's) from temperate
glaciers in the North Cascade mountains of Washington State.
The HC's are distributed throughout the ice, but they tend to
concentrate in thin films on annual late summer ablation
surfaces. Their concentrations are generally on the order of
500 ng/liter of surficial snow, but they may reach 3-5
mg/liter levels in the vicinity of glacial termini or
moraines.

EXPERIMENTAL

Samples have been collected from a number of remote high altitude sites where the concentrations are typically low and from termini where an obvious oiliness or waxiness often exists on the snow surfaces. The samples have been examined by infrared (IR) spectrophotometry, thin layer chromatography (TLC), fluorescence spectrophotometry (FS), and by packed column gas chromatography (GC).

The samples are determined to be complex mixtures of partially-oxidized HC's that in many ways resemble the organic fractions of aerosols (Ketserides et al. 1976), petroleum or coal bitumens, or soils and sediments (Blumer 1976). Some apparent differences exist, however, not only between the glacial HC's and the other naturally-occurring organics, but also between the high and low concentration glacier samples themselves. We have broadly categorized our samples as "high" or "low" depending upon the altitude at which they were obtained. The low samples are invariably richer in HC's.

Typical of this kind of study, major problems exist in acquiring, transporting, and handling samples in such a manner that they are both unmodified and uncontaminated. To avoid contamination, all tools are chemically cleaned with solvents and cleaning solutions and then thermally treated at 600-700°C to either calcine or drive off any residual organic matter. Samples are also subject to loss by evaporation and by container-wall adsorption once ice is melted. Glass and plastic have been found to be particularly unreliable because trace organics readily form "bath tub rings" on their walls. The fragile odor of glacial HC's will disappear within a few hours in glass containers. In metal containers, however, the odor will last 36-48 h. Volatilization is suppressed by working with filled containers whenever possible.

In practice, topical samples are easy to obtain by shaving the top few centimeters of snow and dropping the sample directly into wide-mouth sample cans. Films from previous years are exposed in cross-section in the interior walls of bergschrunds and crevasses, but they are much more difficult to reach. With practice, we have been able to rappel down ice faces and recover samples with 1-3.5 in stainless pipe corers that are hammered up to 6 in into glacial strata. The surfaces of the cores melt slightly and the samples slide easily from the pipe corers into clean, wide-mouth sample cans. The samples, ranging in size from 5-50 liters, can be collected in 1-2 h with diligent effort.

Although problems exist with the techniques, samples are often melted on site and passed immediately through Rohm and Haas XAD-2 or XAD-4 macroreticular resin columns. This preconcentration step has been developed and tested for use in determining trace organics in water samples (Junk et al. 1974) where recovery of insoluble organics is commonly about 90-100% complete. On return to the laboratory, the organics are eluted from the columns with a small volume (25 ml) of diethyl ether. The eluent is reduced in volume to about 0.5 ml by careful evaporation, then the residual solution is used directly for TLC or GC analysis. At this point, many of the organic compounds present are in the low parts per million (ppm) concentration range.

We utilize a two-stage TLC analysis on alumina and acetylated cellulose plates, respectively. Isolated fractions are then examined by FS in microcells and compared with known compound spectra. Our GC work has been carried out on a Varian 1200 GC, used with packed OV-1 or OV-101 partition columns and known calibration samples. Arrangements are being made to extend the analyses to temperature-programmed mass spectrometry (MS) and GC/MS.

RESULTS AND DISCUSSION

We have observed a minimum of 50-60 compounds in our samples. A large portion of them are polar, but a significant fraction of the samples are polycyclic aromatic HC's (PAH's). These compounds are of particular interest since they can be associated with a number of sources (Blumer 1976), and thus can aid in determining the origin of glacial HC's. We have tentatively identified naphthalenes, anthracene, phenanthrene, and benz(a) and benz(e)pyrenes in both the high and low samples. The ratio of PAH to total sample weight appears to be significantly higher in the high altitude samples, but the low samples contain proportionately greater amounts of polar and high molecular weight compounds.

The presence of the particular PAH we have identified strongly indicates that the organics originate from the precipitation fallout of both anthropogenic and biogenic combustion products. This finding is consistent with the fact that snowflakes have been shown to be much more efficient than rain in removing all types of suspended particulate matter from the air (Knutsen et al. 1976, Graedel and Franey 1975). Airborne organics can be expected to concentrate wherever snow accumulates.

Of more specific interest here are the low altitude samples that arise principally from surrounding vegetation. Rasmussen (1970) proposed that plant emissions are a part of a plant-soil-plant carbon cycle in which the heavy terpene vapors are fixed in soils by composting mechanisms. The presence of snow or ice below or around foliage interferes with the cycle and a slow buildup of oxidation products occurs annually on nearby glacial ice. Under ideal conditions, low altitude glacial ice may possibly contain evidence of the sequence of oxidation mechanisms that befall terpenes, from their initial emission to their aging in the presence of air and sunlight as summer months progress.

Based partially on the amount of organic build-up on glacial ice and partially on the preliminary analysis of low altitude organics, we are beginning to suspect that a large portion of the terpenoid material emitted by plants oxidizes right on leaf or needle surfaces. Oxidation products build on plant surfaces until they are either washed off or until they are shaken off by the wind. In the latter event, they might reach the atmosphere in small amounts as aerosols, but by and large, the whole mechanism would essentially preclude large amounts of long-lived biogenic organics in the atmosphere. In order to further investigate this possibility, we are attempting to monitor both the quantity and nature of organics that are building up around selected, isolated treeline vegetation from early spring to late fall in the North Cascades.

REFERENCES

Blumer, K. 1976. Scientific American. 234:35.

Graedel, R. E. and J. P. Franey. 1975. Geophys. Res. Letts. 2:325.

Junk, G. A., J. J. Richerd, M. D. Griesen, D. Witials, J. C. Witiak, M. D. Arguello, R. Vick, H. J. Svec., J. S. Fritz, and G. V. Calder. 1974. J. Chromatog. 99:745.

Ketserides, G., J. Hahns, R. Jaenicke, and C. Junge. 1976. Atmos. Environ. 10:603.

Knutsen, E. O., S. K. Sood, and J. D. Stockham. 1976. Atmos. Environ. 10:395.

Rasmussen, R. A. 1970. Environ. Sci. Technol. 4:4667.

14. CHARACTERIZATION OF THE HAZE IN THE
GREAT SMOKY AND ABASTUMANI MOUNTAINS

Robert K. Stevens, Thomas G. Dzubay, and
Robert W. Shaw, Jr.. Environmental Sciences Research
Laboratory, U.S. Environmental Protection Agency,
Research Triangle Park, North Carolina 27711

ABSTRACT

Field studies were conducted in the Great Smoky
Mountains of Tennessee and the Abastumani Mountains (Adjar-
Imeretian range in Georgian SSR) to measure the composition
of the aerosol that pervades these regions. Sampling was
performed using dichotomous and high-volume samplers and a
variety of instruments to measure gaseous pollutants,
including a gas chromatograph for measuring hydrocarbons.

Sulfate and its associated cations represented 61% of
the fine particle mass in the Smoky Mountains. The average
ionic composition of the sulfate, hydrogen ion, and ammonium
ion was equivalent to ammonium bisulfate. Of the total mass
measured in the fine particles, elemental carbon accounted
for 5% and organic carbon accounted for 10%. Crustal
elements such as aluminum, calcium, iron, and silica were
minor constituents of the fine particle mass. During the
period of this study, the fine particle aerosol in the Great
Smoky Mountains was dominated not by natural organic
compounds but by acid sulfates.

The natural hydrocarbon concentration in the Abastumani
Mountains did not exceed 200 ppbC. The hydrocarbon
concentration attributed to anthropogenic sources ranged from
20 to 400 ppbC.

INTRODUCTION

In 1960, Went suggested that the blue haze often seen in rural areas on sunny days was due to fine particle aerosol produced in the photochemical reactions of natural volatile organic emissions from plants and trees (Went 1960). Since 1960 investigators have discovered that plants and trees do, in fact, emit a variety of volatile organics, and they have concluded that natural organic vapors may contribute significantly to the formation of aerosols (Rasmussen and Went 1965, Zimmerman 1979, 1977). Some controversy still exists concerning the relative contributions of natural and anthropogenic organic compounds to aerosols and decreased visibility. In studies to characterize air masses passing over several rural locations (a pine forest in North Carolina, citrus groves in Florida, and portions of the Everglades), Bufalini and coworkers found that the gaseous samples were dominated by volatile organics normally associated with automotive emissions, rather than natural emissions (Lonneman 1978, Bufalini 1978). These studies may indicate that anthropogenic pollution pervades forested areas in the Eastern United States and overshadows emissions from biogenic sources.

The above studies were mainly concerned with measurements of organic aerosol precursors in the gas phase. More recently, aerosol composition was investigated at a rural site in the Ozark Mountains (Weiss et al. 1977). The investigators concluded that the haze in the Ozark Mountains was caused not by natural organic compounds, but by sulfate aerosol particles. They did not quantitatively measure the composition of the aerosol, but used a nephelometer with variable humidification to infer the presence of various forms of aerosol sulfate. Pierson and others collected aerosols at a remote site on a forested mountain area in Pennsylvania during 17 days in the summer of 1977 (Pierson et al. 1980). The dominant chemical feature of the aerosol was the stoichiometric balance of the sulfate anion ($SO_4^=$), with the cations ammonium ion (NH_4^+) and hydrogen ion (H^+); hydrogen ion was the more abundant cation when sulfate levels exceeded 12 $\mu g/m^3$. Sulfate and its associated hydrogen ion, ammonium ion, and water comprised nearly all of the aerosol mass, with sulfate anion alone comprising as much as 50% of the aerosol mass. The H^+ and $SO_4^=$ levels, with sizes concentrated in the light scattering region (mass median equivalent diameter = 0.84 μm for each species), were proportional to the atmospheric light scattering coefficients b_{scat} (correlation coefficients r = 0.88 and 0.90, respectively), and inversely related to visual range.

Sulfate was not correlated to sulfur dioxide (SO_2) or trace metals. The carbon content of the aerosol was small, averaging only 6% of the $SO_4^=$ concentration.

The studies cited above represent two opposing views of the dominant chemical components of aerosol in rural areas (natural organic vs. sulfate), and consequently suggest the need for a further coordinated effort to characterize both the gases and aerosols in a forested area relatively distant from multiple anthropogenic sources. Thus, as an initial step in addressing this problem, two field studies were conducted to test a variety of new techniques for aerosol characterization and to determine which instrumentation and analysis methods were most useful in understanding the origin and composition of atmospheric haze. The Great Smoky Mountains field study was conducted in the Great Smoky Mountains National Park from September 20 to 26, 1978. The experiment consisted of collection and analysis of aerosols, and gas phase analysis for carbon monoxide (CO), nitric oxide (NO), nitrogen dioxide (NO_2), SO_2, ozone (O_3), halocarbons, sulfur hexafluoride (SF_6), nitric acid (HNO_3), and C_2-C_{10} organic compounds.

Aerosols were collected so that their elemental and ionic composition could be determined. Two additional aerosol experiments were carried out: one to measure gaseous and particulate nitrate in a manner that permits correction for sampling artifacts, and one to determine the extent, if any, of artifact formation on glass fiber filters. Gas measurements were chosen to aid in determining whether the sampled air was representative of natural or anthropogenic sources. The rationale was as follows: (1) halocarbons, CO, NO_x, and sulfur gases indicate intrusion of air from anthropogenic sources; (2) CO and NO_x are associated with emissions from automobiles; (3) halocarbons are expected to be associated with population density (F-11, F-12) and possibly air conditioned automobiles (F-12) but not power plants except where considerable use of refrigeration occurs; and (4) sulfur gases indicate intrusion from fossil fuel combustion activities. Measurements of gaseous organic compounds were made to determine the possible contribution of biogenic hydrocarbon (HC) emissions to aerosols. A second field study was conducted in the Abastumani Mountains in the Soviet Republic of Georgia during July 1979. Gaseous and aerosol measurements were made as in the Smoky Mountain study. In addition, on-site gas chromatographic measurements of volatile gas phase organics were performed. Data from these gas chromatographic studies will be discussed. The data on the aerosol composition will be presented elsewhere.

EXPERIMENTAL PROCEDURE

Smoky Mountains

Measurements were made in a northern portion of the
Great Smoky Mountains National Park near the Elkmont
campground, at 35° 40'N and 83° 36'W. This location is 10 km
southwest of Gatlinburg, Tennessee, and 40 km southeast of
Knoxville, Tennessee. The site elevation at Elkmont is
646 m, and the elevation of mountain peaks located to the
southeast and east is 2000 m. The site is surrounded on all
sides by forest, which is a wilderness for 30 km to the
south.

Measurements were made in the center of a 45 m x 55 m
clearing in the forest. About 30 m to the east was a small
river, and about 60 m to the east was a lightly traveled road
that carried automobile traffic to a camping area located
about 1 km to the south. Although a precise traffic count
was not made during the study, we estimated that, on the
average, fewer than 10 vehicles per hour passed the site.
Equipment at the site included four aerosol samplers and a
mobile laboratory containing instruments to measure gaseous
pollutants. An additional aerosol sampler for collecting
organic compounds and two gas chromatographs for measuring
halocarbons were also operated; the results will be reported
separately (Cronn and Harsch, In preparation). Electrical
power was brought in on a transmission line. Throughout the
study, no vehicles or engines were operated closer than 60 m
to the sampling site.

Gas Measurements

The mobile laboratory was equipped with a gas sampling
manifold and instruments for continuous measurement of SO_2,
O_3, NO, NO_2, and CO. Details of instrumentation used to
monitor these pollutants are given in Table I. All the gas
monitors are commercially available except the CO monitor.
This monitor is based on infrared absorption and was designed
for high sensitivity (20 ppb minimum detectable limit) and
specificity (Burch et al. 1976). Specificity is achieved by
the use of a rotating gas filter cell as described by Chaney
and McClenny (1977). Multipoint calibrations of all
instruments were performed just prior to the study; zero and
span checks were made daily (Stevens et al. 1969, Rehme et
al. 1974). For the CO monitor, calibrations were made at

more frequent intervals to account for residual instrument drift during ambient temperature excursions.

TABLE I. INSTRUMENTS FOR MEASURING GASEOUS POLLUTANTS

Pollutant	Instrument	Method	Procedure
SO_2	Meloy Model 285	Flame Photometric Detector	Permeation Tube[1]
NO and NO_2	Columbia Scientific Instruments Model 1003-AH	Chemiluminescence	Gas Phase Titration[2]
O_3	Bendix Model 8002	Chemiluminescence	Gas Phase Titration[2]
CO	Ford Aerospace	Gas Filter Correlation	Standard Gas Cylinder

[1] From Stevens et al. 1969
[2] From Rehme et al. 1974

Ambient air samples for analysis of volatile organic compounds were collected in 20 1 Tedlar (Dupont 2 mil polyvinylfluoride film) bags which were covered with 5 mil black polyethylene to protect the samples from sunlight during collection and storage. Samples were introduced into the bags by pulling outside air through a clean metal bellows pump coupled to the bag through a Swagelok quick fit connector. After the bags were filled, they were stored in a cardboard box and taken by commercial airlines to Houston, Texas. Analysis was then performed immediately using gas chromatographic procedures described by Seila (1979). The maximum storage time between sampling and analysis was 94 hours.

Aerosol Measurements

To enable a variety of chemical components to be measured, aerosol samples were collected in three separate dichotomous samplers. Two of these were automated dichotomous samplers (A and B) designed for collecting

85

particles in the 0 to 2.4 μm and 2.4 to ~20 μm aerodynamic diameter size ranges (Loo et al. 1976). The filters consisted of highly efficient 1 μm pore size Teflon membranes, 37 mm in diameter, mounted on 5-cm square plastic frames and were obtained from Ghia Corporation, Pleasanton, California. To measure daytime and nighttime conditions, filters in the automated dichotomous samplers were changed at 0630 and 1830 EST. A model 200 dichotomous sampler manufactured by Sierra Instruments, Inc., Carmel Valley, California, was the third sampler (C) used; it collects particles in the 0 to 2.5 μm and 2.5 to ~15 μm ranges. The filters were 37-mm diameter quartz, type 2500 QAST, manufactured by Pallflex Products Corporation, Putnam, Connecticut, and were changed every 24 hours.

Table II presents a summary of aerosol sampling conditions and analyses performed. The flow rates indicated in Table II were measured at the inlet of each sampler using a model S110 dry test meter made by Rockwell Manufacturing Company. The sampling flow rates and volumes are presented in actual conditions; no pressure correction has been made, even though the average barometric pressure at our sampling site was about 7% lower than the value at sea level. The flow fractions listed in Table II represent that portion of the sampled air flow which passed through the coarse particle filter in each dichotomous sampler. Using the known value for the flow fraction and the equations given by Dzubay, data were accurately corrected for fine-coarse mixing effects (Dzubay et al. 1977).

No attempt was made during sampling to prevent the neutralization of collected acid aerosol particles by gaseous ammonia. However, to prevent neutralization by ammonia after sampling, the Teflon filters were removed from the samplers at the end of each sampling period and placed in a glass chamber that contained about 50 g of phosphorus acid (H_3PO_3) crystals. The glass chamber also contained several ammonium sulfate ($[NH_4]_2SO_4$) control samples and sulfuric acid (H_2SO_4) control samples, which were prepared in the laboratory by generating aerosols of pure $(NH_4)_2SO_4$ and H_2SO_4 prior to the field experiment. The quartz filters were stored in 50-mm diameter plastic Petri dishes that were lined with aluminum foil. All samples were stored at ambient temperature (~20°C). Table II indicates the lengths of time that the samples were stored.

TABLE II. SUMMARY OF AEROSOL SAMPLING CONDITIONS AND ANALYSES

Sampler	Filter Medium	Deposit Diameter (mm)	Flow Rate (l/min)	Flow Fraction [1]	Duration (hours)	Analyses	Storage Time (days)
A	Teflon	30.5	52	0.05	12^2	Acidity, Mass, Elemental	3-9 16-22 17-23
B	Teflon	30.5	52	0.05	12^2	H^+, NH_4^+ NO_3^-, $SO_4^=$	6-12
C	Quartz	29.5	16.7	0.10	24^3	Organic carbon, Elemental carbon	110 110

[1]This represents the fraction of the sampled air flow which passes through the coarse particle filter in the dichotomous sampler
[2]Sampling periods began at 0630 and 1830 EST
[3]Sampling periods began at 1830 EST

87

At the end of the field study, the samples were returned to our laboratory. The aerosols collected in sampler A were first analyzed for acidity by exposing the filters to a vapor consisting of ^{14}C-labeled trimethylamine and by counting the beta rays emitted by the ^{14}C retained on the filter (Dzubay et al. 1979). The method was standardized using some of the H_2SO_4 control samples described above. When the beta ray counting was complete, the samples were rendered nonradio-active by exposing them to ammonia vapor for a few minutes. Through an exchange reaction, the ammonia molecules replaced the trimethylamine bound to the sulfate in the samples. Next the samples were analyzed for mass, using beta ray (^{147}Pm) attenuation in an automated instrument developed by Goulding et al. (1978). Because of the prior exposure to ammonia, any sulfate in the samples would be in the form of ammonium sul-fate when the mass analyses were done. The samples were then analyzed for elements with atomic numbers in the range 13 to 56 (aluminum to barium) and also 82 (lead), using an energy dispersive X-ray fluorescence (XRF) spectrometer developed by Jaklevic et al. (1976). The method used for calibration and analysis has been previously published in detail by Stevens et al. (1978) and by Dzubay and Rickel (1978).

Filters representing the fine aerosol fraction collected in sampler B were extracted in 20 ml of aqueous solution using a technique described by Stevens et al. (1978). Hydrogen ion concentration was determined by potentiometric titration on 5 ml of sample extract (Stevens et al. 1978, Brosset and Ferm 1978); ammonium ion concentrations were determined by ion selective electrode on 7 ml of sample extract, and sulfate and nitrate concentrations were determined on 2 ml of sample extract by ion chromatography (Stevens et al. 1978, Sawicki et al. 1978).

Fine fraction aerosols collected on quartz filters in sample C were sent to Oregon Graduate Center where they were analyzed for organic carbon and elemental carbon by a method developed by Johnson and Huntzicker (1979). The reported organic portion is the amount of carbon released in a helium atmosphere at a temperature of 580°C; the elemental portion is the amount of additional carbon released by combustion when oxygen is added. Tests of this method by Johnson and Huntzicker have shown that soot and several types of organic compounds are analyzed with nearly 100% efficiency.

Abastumani Observatory

The Abastumani Astrophysical Observatory is about 20 km west of Tbilisi, the capital of Soviet Georgia, and on the

crest of Mt. Kanobili, 1700 m above sea level. The
surrounding forests are dense and consist mainly of pine and
fir. A small village lies in the valley to the east of the
observatory. This site was chosen for a study of aerosols
because of its distance from industrial areas, the expanses
of coniferous forests which were expected to emit aerosol
precursors, and the ability of the observatory to support an
extended field study.

A Varian gas chromatograph (Model 2440-10) with a
support coated open capillary column and flame ionization
detector was used. Samples were injected through a silicon
rubber septum into a stainless steel cold trap which was
cooled by liquid oxygen (b.p. -183°C). All column and cold
trap hardware were stainless steel. The condensed sample was
subsequently released to the column by warming the trap to
100°C. The column temperature was increased from 30°C to
100°C during the run at a rate of 6°/min.

The carrier gas was nitrogen (Airco, 99.995%); the
detector gases were hydrogen (Airco, 99.999%) and air (Airco,
<1 ppm THC). Carrier gas flow rate was checked daily or more
often. The flow rate was set at 30°C with a reproducibility
of 3% or better; this level of imprecision is thought to be
due to temperature changes in the column.

Three standard gas mixtures were used for system
response calibration; all were contained in stainless steel
vessels. Standard #1 consisted of 14 HC's, ranging from C_2
to C_9 and 1,1,2-trichloroethane in concentrations ranging
from 20 to 100 ppbC in nitrogen at 30 psi. Standard #2 was
isoprene and d-limonene; standard #3 was benzene and
α-pinene. Standards #2 and #3 were at concentrations of
approximately 10 ppmC and in air at 160 psi and 300 psi,
respectively.

Samples were collected in two ways: (1) syringe samples
were taken from the ambient atmosphere, and (2) samples were
taken from a living branch which was partially enclosed for a
short time by the syringe. Nearly all the ambient samples
were taken in the forest about 10 m from the experimental
platform and injected into the chromatograph within 5
minutes. Several ambient samples were taken in the valley
near the Abastumani village and injected within 30 minutes.
Several ambient samples were also taken from a meteorology
tower at about 20 m, just above the pine canopy.

Smoky Mountain Gas Measurements

Table III lists the 24-hour average ambient concentrations of O_3, NO, NO_2, SO_2, and CO measured from the mobile laboratory at the Elkmont monitoring site. Values in parentheses are the maximum values of the distribution of 1-hour averages.

The O_3 concentrations followed a typical photochemical pattern, rising in the midmorning to levels between 40 and 80 ppb, then falling in late afternoon to concentrations below 10 ppb and remaining low through the night. An exception to this pattern occurred on September 21 at 2200 EST, when the O_3 concentration rose to 70 ppb simultaneously with the onset of a thunderstorm. After the storm, the O_3 concentration gradually fell and reached 10 ppb at 0800 EST on September 22.

TABLE III. 24-HOUR AND MAXIMUM 1-HOUR AVERAGES:
ELKMONT SITE, GREAT SMOKY MOUNTAINS

Date	Concentration (ppb)				
	O_3	NO	NO_2	SO_2	CO
9/21/78	38 (70)[1]	--	--	--	<400[2]
9/22/78	37 (70)	<5[3]	<5[3]	3[2] (4)[1]	<400
9/23/78	19 (55)	<8	<11	8 (10)	<370
9/24/78	18 (40)	<6	<7	10 (10)	<170
9/25/78	30 (25)	<4	<4	11 (15)	<190
9/26/78	18 (25)	--	--	15[5] (18)	<190[4]

[1] Values in parentheses are maximum 1-hour averages
[2] 1600-2400 EST
[3] 1800-1400 EST
[4] 0100-0700 EST
[5] 0100-1700 EST

Throughout the study, considerable short-term (several minutes) fluctuations in O_3 concentrations were observed. The fluctuations occurred too rapidly to be explained by variation in cloud cover, and were apparently due to intrusion of air containing reactive species.

The NO and NO_2 concentrations were typically below 5 ppb and showed no long-term trends. A few rare, short-term peaks of several minutes duration, however, ranged as high as 70 ppb and 60 ppb for NO and NO_2, respectively. These peaks generally occurred during the daylight hours and were visually associated with local vehicular traffic. Peaks in NO concentration were correlated with short-term reductions of O_3 concentrations during the day.

Table III shows that 24-hour average levels of CO did not exceed 400 ppb. Carbon monoxide events ranged from 100 to 850 ppb and did not exhibit a typical urban diurnal pattern. Two types of CO events were observed: (1) long-term events lasting a few hours and occurring no more frequently than once a day, and (2) short-term events lasting a few minutes and occurring as many as 100 times a day. We believe that the short-term events were caused by passing automobiles. Long-term CO events occurred between 0900 and 1100 EST on September 22 and between 1800 and 2400 EST on September 23, when concentrations rose from 300 to 850 ppb. The CO events on September 22 were accompanied by increases in Freon 11 and Freon 12 concentrations (Cronn and Harsch, In preparation), but the second CO event did not show a corresponding increase in halocarbon concentration. Hence the two air masses passing through the sampling site during these times had different geographic origins. The daily short-term CO events were clearly anti-correlated with large, short-term decreases in O_3, strongly suggesting that NO from automobiles traveling through the park was responsible. Concentrations of NO were often below detection limits, so that direct comparison of CO with NO was not possible.

Table IV is a summary of the results of volatile organic compound measurements made on September 25 and 26. Details of the individual HC's measured in each period are included in a report by Arnts and Meeks (1980). These investigators concluded that most of the volatile organic compounds measured were emitted by vehicular traffic, with no more than 6% of the total organic mass measured produced by biogenic emissions. The biogenic contribution was estimated by measuring the amount of isoprene and α-pinene present in the samples. In the nine samples, the highest isoprene (which was the dominant natural HC) concentration measured was 6 ppbC.

TABLE IV.   AVERAGE VOLATILE ORGANIC COMPOSITION OF
GREAT SMOKY MOUNTAIN AIR FOR SEPTEMBER 25 and 26, 1978

Organics (Class)	Concentration (ppbC)	
	9/25/78 [1]	9/26/78 [2]
Paraffins	42	55
Olefins	9	17
Aromatics	18	24
Oxygenates	13	13
Unknown [3]	15	16
Acetylene	3	
Total Volatile Organic Compounds	100	132

[1] Averages for the following sampling periods:
600-800, 1000-1020, 1200-1215, 1415-1445, 1600-1620 EST
[2] Averages for the following sampling periods:
825-827, 830-838, 1000-1020, 1130-2235 EST
[3] Unidentified peaks

Smoky Mountain Aerosol Measurements

     Several features are apparent in the results (Table V)
of the average mass and composition of aerosols collected
with dichotomous samplers A and C. First, only about 20% of
the total mass occurs in the coarse fraction, in contrast to
about 36% to 57% previously reported by Stevens for several
urban areas (Stevens et al. 1978). The relatively low coarse
particle mass observed in the Great Smoky Mountains may be
due to factors that tend to minimize the suspension of coarse
particle dust such as the abundance of foliage and the lack
of major anthropogenic activities. A second feature of the
Table V data is that 95% of the sulfur occurs in the fine
fraction, with sulfur accounting for 16% of the fine fraction
mass. We will show below that sulfur and its associated
cations account for 61% of the fine fraction mass. In our
previous study of urban areas, sulfur was not so highly
abundant in the fine fraction (Stevens et al. 1978). The
most surprising feature noted in Table V is the low abundance
of carbon in the aerosol. Organic carbon accounts for only
9% of the fine particle mass. Similar results were obtained
in a study conducted during the summer of 1977 on a forested
mountain in southwestern Pennsylvania (Pierson et al. 1980).

92

Such findings contradict the hypothesis that products of
terpenes derived by photochemical oxidation account for a
major portion of the fine particle aerosol mass in forested
areas.

Table VI shows the ionic composition and sulfur
concentrations measured for fine particles collected by two
different dichotomous samplers. In comparing the sulfur
values measured by XRF with sulfur values from ion
chromatographic (IC) sulfate measurements, one finds that the
correlation coefficient is 0.97, and the ratio of the
averages is 0.94 (XRF/IC). Since the sulfur calibration of
the XRF is accurate to about 8%, the ratio is equal to unity
within experimental error and is consistent with the notion
that sulfur is entirely in the form of sulfate. The data in
Table VI on ionic composition indicate that a negligible
amount of nitrate was collected on the Teflon filters.

Figure 1 is a plot of sulfate versus the sum of the NH_4^+
and H^+ concentrations. The ion concentrations are plotted in
units of nanoequivalents per m^3 (neq/m^3) to enable testing of
the ion charge balance. The linear regression has a 95%
confidence interval for the intercept that overlaps zero and
a 95% confidence interval for the slope that nearly overlaps
unity, which indicates that the sulfate occurs as varying
proportions of $(NH_4)_2SO_4$, NH_4HSO_4, and H_2SO_4.

Figure 2 is a time series plot of H^+ and NH_4^+ molar
concentrations. It is apparent that during the early portion
of the study there is more NH_4^+ than H^+. During the later
portion the reverse is true.

The relative purity of H_2SO_4 in the aerosol can be
estimated from the ratio:

$$f = \frac{[H^+]}{[H^+] + [NH_4^+]}$$

This ratio would have values of 0, 0.5, and 1 for $(NH_4)_2SO_4$,
$(NH_4)HSO_4$, and H_2SO_4, respectively. From the data in Table
VI, the ratio f has a mean value of 0.47 and a standard
deviation of 0.09. The minimum value is 0.34, and the
maximum is 0.62. All of the samples were therefore acidic,
and the average acidity is close to that of $(NH_4)HSO_4$. Such
results pertain to the average composition of all particles
but not necessarily to the composition of individual
particles.

TABLE V. MASS AND ELEMENTAL CONCENTRATIONS FOR AVERAGE DAYTIME, AVERAGE NIGHTTIME, AND AVERAGE FOR ALL SAMPLES COLLECTED AT ELKMONT, TENNESSEE BETWEEN SEPTEMBER 20 and 26, 1978

Entity Measured	Fine (ng/m^3)			Coarse (ng/m^3)		
	Day	Night	Mean	Day	Night	Mean
Mass	26400	22000	24000	6200	4900	5600
C (Organic)			2200			1200
C (Elemental)			1100			<100
Al						
Si	25	15	20	258	200	195
S	48	27	38	762	399	580
Cl	4145	3344	3744	194	214	204
K	<10	<10	<10	9	5	7
Ca	38	42	40	118	98	108
Ti	19	12	16	459	186	322
V	<2	<2	<2	27	9	18
Fe	<2	<2		<2	<2	<2
Ni	29	26	28	166	70	118
Cu	1	1	1	2	2	1
Zn	3	3	3	<3	<3	<5
As	8	10	9	<3	<3	<4
Se	2.5	2	2.2	<1	<1	<1
Br	1.5	2.3	1.4	0.2	0.2	0.2
Pb	11	7	18	7	2	5
	126	68	97	22	6	14

TABLE VI. IONIC COMPOSITION AND SULFUR CONCENTRATIONS
IN THE FINE FRACTION BY TWO DIFFERENT METHODS

Day	Time (EST)	Acidity (neq/m³)	Ionic Composition[1] neq/m³				Sulfur (ng/m³)	
			H^+	NH_4^+	NO_3^-	$SO_4^=$	IC^2	XRF^2
20-21	1830-0630	94	80	143	<6	193	3090	3330
21	0630-1830	112	113	189	<6	333	5330	4400
21-22	1830-0630	105	81	155	<6	227	3630	3740
22	0630-1830	145	169	170	<6	360	5760	4620
22-23	1830-0630	63	58	74	<6	130	2080	2100
23	0630-1830	90	77	88	<6	172	2750	2720
23-24	1830-0630	90	62	107	<6	175	2800	2720
24	0630-1830	90	90	75	<6	172	2750	2810
24-25	1830-0630	95	108	100	<6	211	3380	3450
25	0630-1830	201	196	119	<6	351	5620	5360
25-26	1830-0630	164	158	148	<6	332	5310	4730
26	0630-1830	171	174	152	<6	341	5460	4960
Blanks[3]			0.8 (0.9)	0.3 (2)	<6	0.6 (2)		5 (2)
H_2SO_4 Controls[3]			65 (13)	20 (7)	<6	78 (9)		
H_2SO_4 Controls[3]		105 (57)						1310 (70)
$(NH4)2SO_4$ Controls[3]			3 (3)	-0.1 (0.6)	43 (4)	<6	48 (3)	110 (24)

[1] Ionic composition measured on aerosol collected in sampler B
[2] Acidity and sulfur are measured on aerosol collected in sampler A using amine exposure method and XRF method, respectively
[3] The average values for blanks and controls are computed assuming that the sampling volume is 34 m³. Standard deviations are shown in parentheses

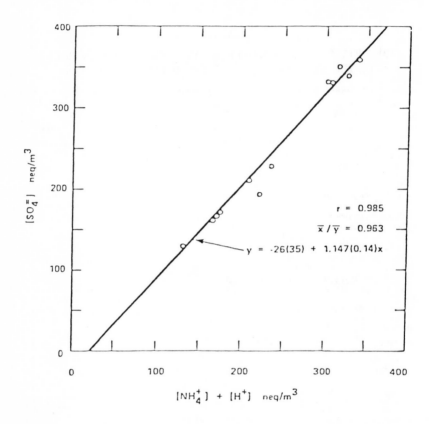

Figure 1. A test of charge balance showing anion vs. cation
concentrations. The numbers in parentheses are
95% confidence intervals for the linear regression
coefficients.

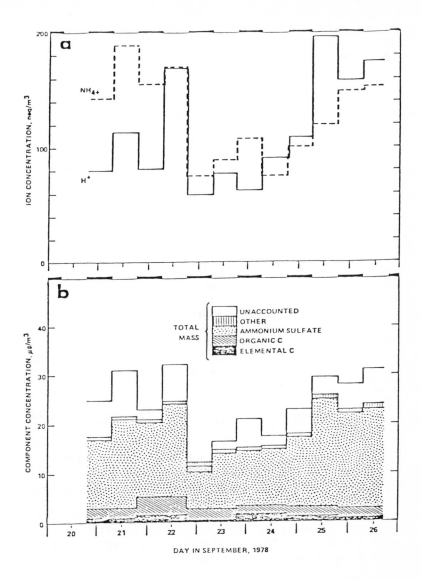

Figure 2. Time series plot for the fine fraction showing:
(1) H^+ and NH_4^+ ionic concentrations in nanoequivalents per m^3 and (2) total fine particle
mass and its various components. The component
labeled "other" includes crustal matter and the
noncarbonaceous portion of vehicle exhaust
aerosol.

How well do these laboratory analyses correspond to the true average acidity of the aerosol particles as they existed in the atmosphere? The data on the ammonium sulfate controls in Table VI show that pure ammonium sulfate standards were not made acidic during storage in the dry, ammonia free chamber. However, the supposedly pure H_2SO_4 controls were about 24% neutralized. In addition, some neutralization of the collected ambient aerosols could have taken place during sampling. In previous studies in which acidity was measured, we used a diffusion denuder at the sampler inlet to remove ammonia during sampling (Dzubay et al. 1979, Stevens et al. 1978). However, no denuder was used for the present measurements in the Great Smoky Mountains. Thus the aerosols could have been even more acidic than the data in Table VI and Figure 2 indicate. However, our acidity values for the Great Smoky Mountains are similar to those obtained by Pierson in a study of aerosols in a forested mountain area (Pierson et al. 1980).

As noted above, sulfate and its associated cations represent the major component of the fine fraction. Ammonium sulfate accounts for 64% of the average fine particle mass measured on the neutralized samples. If one corrects for the known amount of neutralization which occurred in the laboratory, then the original acid sulfate accounts for 61% of the fine particle mass, and organic carbon and elemental carbon only account for 10% and 5%, respectively.

Estimates of Ambient Natural HC Concentrations in Abastumani

Concentrations are calculated based on instrument sensitivities for standards. The sensitivity used for α-pinene was based on iso-propyl benzene, a standard compound with retention times close to that of α-pinene. During the field study, the iso-propyl benzene sensitivity varied with a range of 19% of the mean value. The sensitivity used for isoprene was that of 2,2-dimethyl butane; this sensitivity varied with a range of 32% of the mean.

Observations of HC's lying between α-pinene and d-limonene were made difficult by the presence of a syringe contamination peak. Based on observations of samples collected directly from branches, however, one finds that the ratios of the sum of concentrations of all peaks in the C_{10} region to the concentration of α-pinene has a mean of 3.6 and varies by about 25%. In order to estimate total natural HC's (TNHC's), given in Table VII we have used: [TNHC's] = [isoprene] + 3.6 [α-pinene]. It is important to note that although the values for isoprene and α-pinene are experimentally determined, the value for natural HC is estimated.

Measurements of C_2-C_9 HC's in Abastumani

We use "C_2-C_9" to indicate compounds suspected to be anthropogenic. Concentrations are based on instrument sensitivities for standards. Identification is based on retention time, and hence, is not conclusive. Results are presented in Table VIII.

Haze and Aerosol Acidity

The haze-producing characteristic of an aerosol comes about through the aerosol's fundamental physical properties of refractive index, size distribution, and particle number concentration. The chemical properties of the aerosol are important insofar as they determine the refractive index and the change in size distribution when the aerosol is exposed to varying relative humidity. A laboratory investigation on the characteristic differences in light scattering by aerosols of $(NH_4)_2SO_4$, NH_4HSO_4, and H_2SO_4 under conditions of increasing humidity suggest the importance of the degree of acidity in determining the haze-producing potential of sulfate aerosols (Charlson et al. 1974). In those measurements for $(NH_4)_2SO_4$, there was virtually no increase in light scattering with increasing humidity until values greater than 70% were reached. By contrast, the light scattering by the two more acidic (and hygroscopic) forms of sulfate increased by a factor of 2 to 3 as relative humidity was increased between 50 and 80%.

In the present study a large portion of the fine particle aerosol mass was determined to be composed of NH_4HSO_4 and, possibly, H_2SO_4. Further, the daytime relative humidity was generally above 60%. While it is tempting to speculate that the concurrence of both humidity and acidity may cause an increase in haze over that which would be estimated from the total fine particle mass concentration (Waggoner and Weiss 1980), no quantitative measurements of visibility were performed to examine this possibility. It is important to note, however, that in a field measurement in which relative humidity decreases as well as increases, the phenomenon of hysteresis may lessen the difference between the light scattering properties of the two classes of sulfate aerosols (Tang 1980). When the relative humidity to which a droplet of pure $(NH_4)_2SO_4$ is exposed is gradually decreased below its deliquescence humidity (80%), the water content of the droplet decreases, but not abruptly. At a given humidity (between the deliquescence humidity and the much lower "crystallization" humidity), the water content of the droplet remains well above the value it would have had if the humidity were instead increasing from a very low value. It is quite common for ambient humidity to cycle above and below 80% relative humidity with a diurnal variation, so that ambient $(NH_4)_2SO_4$ aerosol may well exist more often in a state of high water content (and consequently with enhanced light scattering ability) than with the low water content appropriate to the results of Charlson et al. (1974).

In passing we note that Husar and Patterson have analyzed trends in visibility in the eastern United States between 1948 and 1974 (Husar and Patterson 1980). They report a large decrease in visibility for the summer months, July through September, and find that this trend coincides with a trend toward increased coal combustion during summer months. For the region surrounding the Smoky Mountains, the decrease in summertime visibility was very substantial.

A final qualitative observation was our perception of the visible haze as white during the period of our study, independent of viewing angle. There was no indication of the blue haze which Went described previously (Went 1960).

While the present study does not provide any quantitative results relating aerosol acidity and visibility, it is hoped that the high acidity levels which have been measured in the study will encourage future investigation of its effects on visibility.

CONCLUSIONS

Organic compounds, and inorganic pollutant gases associated with industrial and automotive emissions, dominated the gaseous pollutants in the atmosphere. For the period when organics were measured, volatile gas phase organics did not exceed 132 ppb carbon, and biogenic emissions contributed less than 6 ppb carbon to the organics measured.

TABLE VII.   AMBIENT GASEOUS NATURAL HC (ppbC):
ABASTUMANI, JULY 1979

Date	Time (EST)	Site	α-pinene	Isoprene	TNHC[*]
10	1700		<12	--	<12
11	1310		20	27	99
	1600		14	10	60
	1735		14	2	52
13	0745		2	4	11
	1255		6	--	22
	1400		<2	--	<7
	1600		10	--	36
	2200		10	2	38
14	1120		<2	4	<11
	1230		<6	4	<26
	1700		12	9	52
	1800		6	18	40
	2110		12	3	46
	2345		14	5	55
15	0810		8	6	35
	1200		12	--	43
	1600		8	29	58
	2015		14	12	62
16	1015		2	5	12
	1445		6	24	46
	1700		12	8	51
	2200		20	8	80
17	0815	A	26	--	94
	1040		4	7	21
	1210	C	12	73	116
18	0900	B	--	--	--
	1000				
	1100	D	6	2	24
	1530		15		
	1930		10	4	40
	2245		9	--	32

(Continued on next page)

101

TABLE VII.   (Continued)

Date	Time (EST)	Site	α-pinene	Isoprene	TNHC
19	0815	A	3	--	11
	1030		--	--	
	1645		--	--	
20	0945		<6	--	<22
	1230		<10	5	<41
	1515		--	--	--
	1715	E	--	--	--
21	1030		--	2	2
	1200		--	--	
	1400			5	5
	1600		--	33	33
22	0910		15	2	56
	1300		9	28	60
	1400		6	11	33
23	1645		18	3	68
	1730		18	8	73
24	1230	B	--	--	--
	1320		6	--	22
	1620		12	--	43
25	0645	A	54	--	194
	0840	B	6	--	22
26	1200		--	8	8
27	1630		9	14	47

[*]TNHC = [isoprene] + 3.6 [α-pinene]
Sites: All sites not otherwise marked are near the platform

A-Refers to the floor of the Abastumani Valley, near the
town: 7/17 and 7/19 100 m from the funicular station in
fir and pine grove; 7/25, near the radiosonde launch site
B-Refers to 20 m above ground on the meteorology tower, atop
Mt. Kanobili
C-Was very near a fir branch
D-Was very near a pine branch
E-Was deep in forest, ∼0.5 km from platform

TABLE VIII. AMBIENT GASEOUS C_2-C_9 HC (ppbC):
ABASTUMANI, JULY 1979

Date	Time (EST)	Site	C_2	C_3	C_4	C_5^*	C_6	C_7	C_8	C_9	Total
10	1700		--	--	--	--	120	--	--	--	120
11	1310		--	10	18	28	53	6	5	--	120
	1600		--	2	5	16	103	11	3	--	140
	1735		--	3	9	11	98	9	5	--	135
13	0745		--	3	2	4	20	2	4	--	35
	1255		--	8	3	7	38	4	12	--	72
	1400		--	9	3	2	42	5	3	--	64
	1600		--	5	4	6	41	6	4	--	64
	2200		--	6	5	4	41	8	4	--	68
14	1120		--	5	5	2	8	1	2	--	23
	1230		--	4	4	2	29	2	3	--	44
	1700		--	4	4	4	5	--	2	--	19
	1800		--	7	7	8	66	5	4	--	97
	2110		--	5	5	3	21	4	--	--	38
	2345		8	2	3	5	17	--	--	--	35
15	0810		--	4	4	6	3	--	--	--	17
	1200		--	7	1	10	--	--	--	--	18
	1600		--	3	4	27	15	--	--	--	49
	2015		--	5	4	11	21	--	--	--	41
16	1015		--	--	3	17	14	--	--	--	34
	1445		--	8	9	24	17	4	--	--	65
	1700		--	18	7	8	59	4	14	--	110
	2200		--	9	11	13	33	2	7	--	75
17	0815	A	12	3	44	--	281	54	14	--	408
	1040		4	1	10	--	28	3	8	--	54
	1210	C	--	2	8	32	19	--	--	--	61
18	0900	B	--	--	14	--	120	6	19	--	159
	1000		--	3	7	--	22	--	--	--	32
	1100	D	--	2	7	2	6	--	--	--	17
	1530		--	--	4	6	29	10	21	--	70
	1930		--	--	5	2	17	3	7	--	34
	2245		--	2	4	1	9	4	4	--	24

(Continued on next page)

TABLE VIII. (Continued)

Date	Time (EST)	Site	C_2	C_3	C_4	C_5^*	C_6	C_7	C_8	C_9	Total
19	0815	A	--	4	6	1	31	6	17	--	65
	1030		--	3	1	--	27	--	6	--	37
	1645		--	--	--	--	13	--	--	--	13
20	0945		--	--	1	--	62	5	12	--	80
	1230		--	2	1	2	27	--	--	--	32
	1515		--	--	--	--	13	3	--	11	27
	1715	E	--	--	3	1	32	8	19	--	63
21	1030		--	--	6	2	5	3	9	--	25
	1200		--	7	5	--	9	3	9	--	33
	1400		--	6	4	--	15	3	11	--	39
	1600		--	10	19	2	16	3	8	--	58
22	0910		--	18	21	2	8	9	13	--	71
	1300		--	10	14	4	34	3	8	--	73
	1400		--	6	5	--	28	3	7	--	49
23	1645		--	15	24	3	48	--	--	--	90
	1730		--	--	12	--	67	2	3	--	84
24	1230	B	--	4	--	4	27	3	5	--	43
	1320		--	10	2	1	40	2	8	--	63
	1620		--	8	1	1	1	6	--	--	17
25	0645	A	--	8	3	5	57	46	28	--	147
	0840	B	--	8	--	7	41	5	9	--	70
26	1200		--	5	12	4	8	4	9	--	42
27	1630		--	--	12	--	11	--	4	--	27

*Excludes isoprene

Aerosol measurements made in the Great Smoky Mountains revealed that sulfate and its associated cations represented 61% of the fine particle mass. The average ionic composition of the $SO_4^=$, H^+, and NH_4^+ was equivalent to ammonium bisulfate. Of the total mass measured in fine particles, elemental carbon accounted for 5% and organic carbon accounted for 10%. Crustal elements such as aluminum, calcium, iron, and silicon were minor constituents of the fine particle mass. This work shows that during the period of the study the fine particle aerosol in the Great Smoky Mountains was dominated not by natural organic compounds but by acid sulfates.

The gas-phase HC measurements made in Abastumani indicate that, as expected, the C_2-C_9 HC's were highest when the wind direction was from a nearby village. The largest concentration observed was approximately 400 ppbC with 70% found in the region of the chromatograms where C_6 compounds appear. When the wind came from the forests, the C_2-C_9 concentrations often dropped below 20 ppbC. Estimates of natural HC's based on measurements of isoprene and α-pinene ranged from nearly zero to almost 200 ppbC. The maximum values were observed during periods of relatively high ambient temperatures. There was no correlation between C_2-C_9 and natural HC values.

ACKNOWLEDGMENTS

We thank the many people who contributed to the Smoky Mountain Study. Dr. James Huntzicker of Oregon Graduate Center provided the carbon analyses; Dr. Gary Larsen and the National Park Service provided the sampling site and cooperation during the monitoring program; Sarah Meeks and Robert Arnts of EPA carried out sampling and analysis of volatile organic compounds. Carol Sawicki of EPA provided carbon analyses. Northrop Services, Inc. personnel who carried out aerosol analyses were Dr. William Courtney, Chris Presley, Carolyn Owen, Brenda Mullen, Ed Tew, Judy Hunt, and Mark Mason. Dr. Rick Varcoe of Northrop and Mr. Lowell Hines of EPA helped prepare for and support the study.

Work in the USSR was supported by many Soviet scientists. We especially thank Dr. G. Rosenberg, Dr. V. Stepanenko, and Academician E. Kharadze. We also thank Dr. William Lonneman of EPA for his helpful advice and HC standards and Arthur Coleman for terpene standards.

We thank Drs. Herbert Wiser, Basil Dimitriades, William Wilson, William McClenny and Charles Lewis for their participation in the planning and carrying out of these studies.

REFERENCES

Arnts, R. R. and S. A. Meeks. 1980. Biogenic Hydrocarbon Contribution to the Ambient Air of Selected Areas. EPA/600-3-80-023. U.S. Environmental Protection Agency, January.

Brosset, C. and M. Ferm. 1978. Atmos. Environ. 12:909.

Bufalini, J. J. 1978. Ozone formation from rural hydrocarbons. Proceedings CRC Air Pollution Research Symposium, New Orleans, Louisiana, May.

Burch, D. E., F. J. Gates, and J. D. Pembrook. 1976. Ambient CO Monitor. Final Report, prepared under EPA Contract No. 68-02-2219, Ford Aerospace and Communications Corp., Newport Beach, California.

Chaney, L. W. and W. A. McClenny. 1977. Environ. Sci. Tech. 11:1186.

Charlson, R. J., A. H. Vanderpohl, D. S. Covert, A. P. Waggoner, and N. C. Ahlquist. 1974. Atmos. Environ. 8:1257.

Cronn, D. and D. Harsch. Report in preparation.

Dzubay, T. G. and D. G. Rickel. 1978. Report in Electron Microscopy and X-Ray Applications. Ann Arbor Science, Ann Arbor, Michigan. pp. 3-20.

Dzubay, T. G., G. K. Snyder, D. J. Reutter, and R. K. Stevens. 1979. Atmos. Environ. 13:1209.

Dzubay, T. G., R. K. Stevens, and C. M. Peterson. 1977. In: X-Ray Fluorescence Analysis of Environmental Samples. T. G. Dzubay, ed., Ann Arbor Science, Ann Arbor, Michigan. pp. 95-105.

Goulding, F. S., J. M. Jaklevic, and B. W. Loo. 1978. Aerosol Analysis for the Regional Air Pollution Study: Interim Report. EPA-600/4-78-034. U.S. Environmental Protection Agency, July.

Husar, R. B. and D. E. Patterson. 1980. Regional scale air pollution sources and effects. Annals of New York Acad. Sci. 338:399.

Jaklevic, J. M., D. A. Landis, and F. S. Goulding. 1976. In: Advances in X-Ray Analysis. R. W. Gould, C. S. Barrett, J. B. Newkirk, and C. O. Rudd, eds. 19:253-265, Kendall Hunt, Dubuque, Iowa.

Johnson, R. L. and J. J. Huntzicker. 1979. In: Proceedings, Carbonaceous Particles in the Atmosphere. T. Hovakov, ed., pp. 10-12. Lawrence Berkeley Laboratory Report No. LBL-9037, CONF-7803101, UC-11, Berkeley, California.

Lonneman, W. A., R. L. Seila, and J. J. Bufalini. 1978. Environ. Sci. Technol. 12:459.

Loo, B. W., J. M. Jaklevic, and F. S. Goulding. 1976. In: Fine Particles. B.Y.H. Liu, ed., Academic Press, New York, New York, pp. 312-350.

Pierson, W. R., W. W. Brachaczek, T. J. Truex, J. W. Butler, and T. J. Korniski. 1980. Ambient sulfate measurements on Allegheny Mountain and the question of atmospheric sulfate in the Northeastern United States. Annals of N.Y. Acad. Sci. 338:145.

Rasmussen, R. A. and F. W. Went. 1965. Proc. Nat. Acad. Sci. 53:215.

Rehme, K. A., B. E. Martin, and J. A. Hodgeson. 1974. Tentative Method for the Calibration of Nitric Oxide, Nitrogen Dioxide and Ozone Analyzers by Gas Phase Titration. EPA-R2-73-246. U.S. Environmental Protection Agency, March.

Sawicki, E., J. D. Mulik, and E. Wittgenstein. 1978. Ion Chromatographic Analysis of Environmental Pollutants. Ann Arbor Science, Ann Arbor, Michigan.

Seila, R. 1979. Non-Urban Hydrocarbon Concentrations in Ambient Air North of Houston, Texas. EPA-600/3-79-010. U.S. Environmental Protection Agency, February.

Stevens, R. K., A. E. O'Keeffe, and G. C. Ortman. 1969. Environ. Sci. Technol. 3:652.

Stevens, R. K., T. G. Dzubay, G. Russwurm, and D. G. Rickel. 1978. Atmos. Environ. 12:55.

Tang, I. N. 1980. In: Generation of Aerosols and Facilities for Exposure Experiment. K. Willeke, ed., Ann Arbor Science, Ann Arbor, Michigan, p. 153.

Waggoner, A. P. and R. E. Weiss. In press. Atmos. Environ. 14:623.

Weiss, R. E., A. P. Waggoner, R. J. Charlson, and N. C. Ahlquist. 1977. Science. 195:979.

Went, F. W. 1960. Nature. 187:641.

Zimmerman, P. R. 1977. Procedures for conducting hydrocarbon emission inventories of biogenic sources and some results of recent investigations. Presented at EPA Emission Inventory/Factor Workshop, Raleigh, North Carolina, September.

Zimmerman, P. R. 1979. Tampa Bay Area Photochemical Oxidant Study. Final Report Appendix C. EPA-904/9-77-028. U.S. Environmental Protection Agency, February.

DISCUSSION OF PRESENTATION

LUDLUM: Were you able to identify any of those peaks at the lower mass number?

STEVENS: Yes, they were C_2, C_3, C_4 hydrocarbons. We attempted to make the analysis as simple as possible.

FERMAN: I had a couple of questions and comments on some of the work you did in the Smokies with aerosols. I didn't follow everything you said about color, but several people in the last few years have pointed out that not only is the size distribution of the aerosol important to color, but also the angle to the sun. The same aerosol, if you look at it with front scattering or back scattering, might be dark or white, depending on where the sun is.

STEVENS: Whether we looked at it from above, from across, or with the sun behind us, I couldn't honestly say that there was a marked difference in the texture of the haze. The whiteness may have gotten brighter, or less, or more diffuse.

If you notice, from the airplane shot, looking at it straight down, there was a whitish color. Looking at it from across, it was whitish. With the sun to my right and later on with it to my left, it was still white.

We are talking about a hygroscopic aerosol that has a particle size distribution which ranges from about 0.1 μm to 1 μm, with a mass median diameter possibly somewhere around 0.4 μm to 0.5 μm. With that broad a range, you would expect it to be white. There was no evidence of any significant mass volume below a tenth of a micrometer.

Obviously, there was probably some naturally occurring haze back in our great-great grandfathers' time due to natural HC aerosol, and the haze was possibly blue. Sulfates did not pervade the mountains then. Now, the natural HC aerosol is overshadowed by the aerosol sulfate that invades the area from a variety of locations. We examined wind trajectories for the six-day period. On any given day, the air mass would be coming from the Ohio River Basin, from Birmingham, from Atlanta, North Carolina, or eastern Tennessee.

We observed that the acidity was higher in the daytime than at night, again consistent with the observation that sulfate is being photochemically generated and forming sulfuric acid.

From the ammonia measurements made there, we can say that the ammonia concentrations were less than 1 ppb.

FERMAN: Did you measure nitrate; could you give us some numbers?

STEVENS: Nitrate concentrations were below our detection limit, which was approximately 0.2 $\mu g/m^3$.

FERMAN: With that much acidity on the sulfate, what about the problem of, say, nitrate actually being replaced by sulfuric acid and volatilizing it as HNO_3 off the filter.

STEVENS: That was called the denuder difference experiment, which Bob Shaw performed. He observed that that was not occurring, in that particular aerosol. We have seen it occur in Los Angeles. Bob confirmed that there is what we call a negative artifact. A problem is caused by the dissociation of ammonium nitrate to ammonia and nitric acid. But we did not see evidence of significant nitrate artifact in the Smoky Mountains.

I think we didn't see this artifact because there was an absence of nitric acid; the NO_2 concentrations were typically less than 2 ppb. And so, we can say that what photochemistry was going on appeared to be dominated by the sulfur photochemistry.

FERMAN: What was your SO_2 concentration?

STEVENS: SO_2 concentrations ranged from as low as 2 ppb, to as high as 17 ppb. Those numbers were about the same aloft.

OLLISON: You said the volatile carbon constituents in the aerosol were on the order of several percent. Was that sufficient to account for the presumed estimated emissions from that--organic emissions?

STEVENS: Well, since it is small, and since the total nonmethane HC's were small, it is in the right direction.

OLLISON: So one percent of your mass loading, you think, would account for the emissions from the trees?

STEVENS: No. First of all, the highest number we saw for biogenic emissions was 6 ppbC.

OLLISON: That was your measured concentration, not the estimated emission?

STEVENS: We did not estimate emissions.

OLLISON: Yes, but given what went on the few days before, do you feel that their projected emissions would be accounted for by your 1% or 2% organic constituency?

STEVENS: I don't think we know how much of the photochemical products would end up on the aerosols. Alpha-pinene oxidation products might eventually condense to an aerosol particle and turn up on the filter. My guess is that this mechanism represents a small portion of the carbon fraction we were talking about, because the nonmethane HC fraction associated with auto exhaust was better than 50 to 80% of the HC's. The photochemical products of those materials may more likely be the source of the volatile organic component of the condensed aerosols.

The elemental carbon is probably associated with one or more of several sources: diesels, combustion products from automobiles, natural gas combustion, and coal fired power plants. Finally, almost every aerosol sample we have analyzed contains both sulfur and elemental carbon.

We also cannot rule out that some aerosols may be due to campfires located nearby. We have no evidence at present.

BUFALINI: This afternoon there will be at least a couple of papers presented on the products that are produced from the photooxidation of biogenic HC's. What I specifically have

reference to is some work that Murray Kaiserman will be
presenting, where we could not account for the carbon. We
doubt very seriously it went to the wall. What I'm
suggesting is that you missed a large fraction of the HC that
would have been present in the aerosol, largely because, as
it was collected, its vapor pressure was sufficiently high
for it to have been revolatilized.

STEVENS: It's already in equilibrium; why should equilibrium
change after it was collected? If you put benzoic acid on
the filter, for example, it will stay benzoic acid on the
filter. Even with anthracene, a compound to which this could
happen, it would take days to revolatilize. Samples were
kept in small closed containers to minimize evaporative loss.

I do not, however, rule that out. It is possible that
particulates have a finite vapor pressure. Work recently
published by van der Waag, I believe, did show that for high
volume samples there was a finite vapor pressure and also an
equilibrium between a variety of organic compounds in the gas
phase and the aerosol phase. But he took those samples in an
extremely polluted situation.

BUFALINI: Sounds like a good answer. Before we go into it,
Jack [Durham], you are an aerosol specialist; do you agree
with that?

DURHAM: Well, I'll be surprised, first of all, if there will
not be volatiles that are missed by this technique. Did you
use a heated sampler to collect the sample?

STEVENS: No.

DURHAM: And what was the temperature of that sampler?

STEVENS: Ambient temperature. It averaged around 70°F,
70-75°F.

DURHAM: And what was the duration of storage and transport
time from sampler to the analysis?

STEVENS: Several weeks.

DURHAM: Then I'd say the question is open. There could be
volatile loss.

STEVENS: I don't think there is any question that there is
some volatilization. But whether that is a significant
amount, really, I think is doubtful. It certainly can't
exceed the total amount of volatile organics that we measure.

111

WALKER: It seems to me that the essence of the Smoky mountain problem has always been, not that there is haze there, but why is there haze there when the haze isn't in the plains, the East, or the West, particularly?

STEVENS: Haven't you ever flown over the East Coast?

WALKER: Oh, yes. But in general, it is long recognized that the haze is very intense in the mountains. So, the question is what is different about the mountains? Your observation that there is a lot of sulfate now doesn't change the picture totally. It isn't questioned. But the question is, why was there haze there 50 years ago? I have always felt the most logical explanation for this was simply that the mountains are a mixer. They cause upper layers that contain virgin O_3 to mix with the biogenic material from the surface boundary layer and create the haze. Probably today, with SO_2 being present, it contributes in an inversion; but there's probably nothing in what you've seen that changes that picture. Or, would you agree that is a probable picture?

STEVENS: As you say, it is a receptor. I disagree with your statement that it is more hazy in the mountains. We perceive the haze more readily when viewing distant mountains. This haze is less conspicuous in flat terrain.

WALKER: I am saying that all you have to have is mixing. The O_3 is in the upper layers, anywhere between 20 and 50 ppb, to 80 ppb, all the time.

STEVENS: How do you handle this in the wintertime?

LUDLUM: It is there in the winter.

WALKER: It is there in the winter. It's highest of all in April.

STEVENS: I think there was a potential for photochemical, or O_3, intrusions, as you're saying about 100 years ago. But the haze was blue, not white. These intrusions are probably still happening today, but they are insignificant. They are not significant compared to what is happening with the sulfate.

An interesting thing to do would be to go back and interview the people in the area. They have old photographs and paintings of the area. You could come up with a chronological pattern of how the haze has changed over the years from those.

RASMUSSEN: That has already been done. One thing that came out of that work was that we have lost the effects of wood smoke. However, wood smoke is coming back. In the 1800's throughout the Midwest and all the far West, a lot of wood smoke was generated from the burning, the slash clearing of the grass. There are even little articles in some of the journals of the historical society in Salem, Oregon, about the Willamette Valley. There is a problem now with tremendous wood smoke in the Willamette, Medford, and Portland, Oregon, areas. Medford, Oregon, is the most polluted place in the United States, with a tremendous accumulation of wood smoke.

RASMUSSEN: You have to go back even further in time, I think, to find blue haze. You have to really go back to when there was not a predominance of clearing the land and wood smoke.

STEVENS: I think there is a blue haze in the Abastumani Mountains, and I think that may well be due to nonanthropogenic activity. However, I would reserve that. I think Dr. William Wilson and I would take the position of: "Lets look at the data first." We can only say that we observed a blue haze on certain days and white haze on other days.

W. E. WILSON: I would like to mention that there are drawings from the 11th century of the mountains in Abastumani which show blue haze. This blue haze has been known for at least a thousand years. It's a very unpopulated country. The Russians think the blue haze is very healthful. People go there to recuperate. And they're concerned about losing it.

But one of the reasons you wanted to go to Abastumani was because it is a pristine area.

VOICE FROM AUDIENCE: When did the Smoky Mountains get their name? They must have been named long before the settlers entered the area.

W. E. WILSON: The original name that the Indians had was not the Smoky Mountains, but the Mountains of Many Smokes.

SAME VOICE FROM AUDIENCE: So, in other words, there was a lot of haze there.

W. E. WILSON: Going back to some of these old paintings of the Indian villages, you can actually see the smoke going up, and then going right along the line. There were inversion layers back then; the smoke was a contributing factor.

STEVENS: The Smoky Mountains are contiguous with the Blue Ridge Mountains. As William pointed out, the Indian inhabitants associated the Smoky Mountains with localized events of smoke. The name did not pertain to the general atmosphere that pervaded the whole, overall area.

ARNTS: Maybe I missed something in your reasoning. Is there some reason why you would rule out the possibility that the sulfates may actually be biogenic and may have been there all along?

STEVENS: We used a device to permit us to look to see if there was any H_2S. We have made measurements in a number of locations, and there is very little H_2S emitted, except at spots in and near marshes.

Secondly, the vertical profile of the SO_2 would show much higher concentrations near the ground level if the SO_2 were local. They were more or less ubiquitous, the same concentrations of SO_2 were all around. If there were localized H_2S or SO_2 sources, we would have seen them.

We had a relatively uniform SO_2 concentration on that side of the mountains. The SO_2 is transported inward from other sources.

KELLY: Why wouldn't you have seen any nitric acid? If your NO_2 and SO_2 are about similar, and if your SO_2 is made into sulfuric acid by photochemistry (presumably OH), why would you see nitric acid levels that were even higher, since NO_2 reacts much faster than SO_2 with OH?

STEVENS: The NO_2 concentrations were at our detection level; whereas, the SO_2 concentrations were always above it. Probably since there were a few cars going through the area, some of that NO_2 could have been localized material.

Finally, the nitric acid was in the gas phase. There are many, many mechanisms for the removal of NO_2. I'm not saying it wasn't there; it was just at a level below 2-4 parts per billion. That is all I'm saying.

VOICE FROM AUDIENCE: It wasn't incorporated in the aerosol?

STEVENS: It was below 0.2 $\mu g/m^3$ in the aerosol phase.

That is not inconsistent with numerous measurements made in a variety of locations. So those data are not inconsistent, both in urban as well as in rural areas.

VOICE FROM AUDIENCE: What about your aerosol measurements in Russia?

STEVENS: We are not prepared to make comments on that at the present time. I'll say this, the aerosol mass measurements were perhaps a factor of 5 lower.

W. E. WILSON: How about the carbon in it?

STEVENS: We also collected samples there in which we are doing carbon dating. Again, those data are not ready for release.

OLLISON: But they're fairly high carbon content?

STEVENS: I can't say.

OLLISON: You have no data on them?

STEVENS: We have no data. Perhaps, in about six to eight months we might have something.

WALKER: You say you are doing carbon dating on the Russian samples. Did you do carbon dating on the Smoky Mountain samples?

STEVENS: No. I wish we had. There is only so much we can do.

15. PHOTOCHEMICAL OXIDANT POTENTIAL
OF THE BIOGENIC HYDROCARBONS

Robert R. Arnts, Bruce W. Gay, and Joseph J. Bufalini.
Environmental Sciences Research Laboratory, U.S.
Environmental Protection Agency, Research Triangle Park,
North Carolina 27711

ABSTRACT

Laboratory experiments demonstrate that isoprene and
monoterpenes are inefficient ozone precursors relative to
propylene in the hydrocarbon/nitrogen oxides photochemical
system. Their inefficiency increases with increasing carbon
to nitrogen oxides ratio. An n-butane/propylene
photochemical model is used to predict maximum concentrations
of ozone that could be produced from biogenic hydrocarbons.

INTRODUCTION

This paper focuses on the fate of the biogenic
hydrocarbons (HC's) and their role in atmospheric chemistry.

A review of the literature reveals that the bulk of the
biogenic HC's are isoprene or monoterpene in character. The
emissions most frequently mentioned are α-pinene from
evergreen foliage and isoprene from deciduous foliage, with
some vegetation emitting both. Other HC's, however, have
also been observed volatilizing from plants; some have been
seen in the essential oils of vegetation and thus might be
expected to be released to the atmosphere. A cross section
of these compounds demonstrates the structural variety
observed (Figure 1).

Biogenic HC's differ from auto-related HC's in several respects. First, the biogenic HC's are of fairly high molecular weight (C_{10}'s), having one, two, or three doubly-bonded carbons (exclusive of the saturated oxygenates camphor and 1,8-cineole). Secondly, in many cases oxidative cleavage of double bonds in cyclic structures can lead to the formation of di-oxygenated species, e.g., cleavage of the ring double bond in α-pinene to form a keto-aldehyde. Such compounds are expected to have significantly lower vapor pressures than their precursor terpene; thus, their tendency to form aerosols is stronger than that of most auto-related lighter HC's.

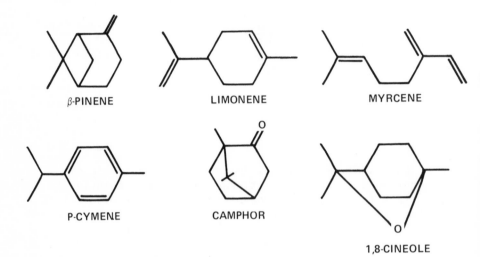

β-PINENE   LIMONENE   MYRCENE

P-CYMENE   CAMPHOR   1,8-CINEOLE

Figure 1.   Volatile biogenic HC's.

A review of HC reactivity studies reveals that isoprene and a few terpenes are moderately to extremely reactive when compared with typical auto-related HC's (Arnts and Gay 1979). "Reactivity" has traditionally been used to rate a HC's ability to yield ozone (O_3) in the presence of sunlight and nitrogen oxides (NO_x). However, most studies have used surrogate parameters to define this ability because of shorter reaction times. These surrogate parameters have included HC loss rate, nitric oxide (NO) loss rate, and more recently the rate of attack by hydroxyl radicals (Glasson and Tuesday 1970, Darnall et al. 1976).

Other studies have specifically addressed the aerosol formation of terpenes under simulated atmospheric conditions (Schwartz 1974, Schuetzle and Rasmussen 1978, Lipeles et al.

118

1978). Quantitative conversion of terpenes to aerosols has
been reported. Attempts to identify specific terpene
photooxidation products have met with limited success; to
date the bulk of the products are partially characterized but
are for the most part unidentified.

This study was initiated to determine the ability of the
biogenic HC's to form O_3 in the NO_x photochemical system and
to determine products formed therein. The specific question
addressed is whether the terpenes generate O_3 in the NO_x
photochemical system as indicated by previous reactivity
data, or whether aerosol production prevents O_3 buildup by
effectively removing the HC from the photochemical chain.

RATIONALE

Smog chambers have been used for years to study the
photochemistry of various HC/NO_x mixtures. These chambers
have been valuable tools for defining the behavior of various
compounds and their products in the atmosphere. However,
because of experimental limitations, smog chamber studies
cannot directly provide quantitative descriptions of the real
atmosphere. Probably the most significant of these
limitations is the influence of the reactor walls in the
photochemical process. In addition, contamination of the gas
matrix by compounds outgassing from or off of the walls,
especially Teflon surfaces, forces the use of higher than
atmospheric concentrations in these chambers. Also, the use
of smog chambers usually prohibits large sample collection –
a necessity when analyzing low levels of HC's.

In this study, the biogenic HC's were studied at higher
than atmospheric concentrations over a range of carbon to NO_x
ratios representative of rural atmospheres. Propylene was
included as a bench mark compound; propylene has been well
characterized in previous chamber studies and its reactions
well modeled. Thus the behavior of the biogenic HC's could
be studied relative to propylene over a range of conditions.

EXPERIMENTAL

To study the reactivity of the HC's with respect to O_3
production, irradiations were conducted until the O_3 maximum
was achieved. These studies were conducted in 250 l
fluorinated ethylene propylene (FEP) Teflon bags in a
thermostatted (25°C) irradiation chamber (k_d = 0.45 min^{-1}).

Initial HC concentration ranged from 0.33 ppmC to 60 ppmC at a constant initial NO concentration of 0.33 ppm. This yielded a C:NO_x ratio ranging from 1:1 to 200:1; the lower end of this range (1:1 to 30:1) encompasses the reported optimum ratio for O_3 production (Coffey and Westberg 1977). The high range is more representative of rural and remote clean airsheds (200:1).

The bag irradiations were monitored using a variety of techniques. Ozone and nitrogen oxides were monitored via chemiluminescence, peroxyacetyl nitrate (PAN) by electron capture detection gas chromatography, and the reactant HC by flame ionization detection gas chromatography. Nitrogen dioxide was determined with the Saltzman method as a check against the interference prone NO_x chemiluminescence monitor. Formaldehyde was also spot checked using the chromotropic acid method.

Supplementary product identification studies were conducted using a long path infrared (IR) Fourier transform spectrometer. The system previously described is a combination absorption cell and irradiation chamber (Hanst et al. 1973). Mixtures of 80 ppmC HC and 1 ppm nitrogen dioxide (NO_2) were irradiated for 60 min and their products monitored. The concentrations used were higher than those used in the bag study in order to satisfy the sensitivity requirements of the IR spectrometer.

RESULTS

Isoprene, p-cymene, α-pinene, d-limonene, and propylene were studied over a range of C:NO_x ratios with respect to maximum O_3 produced. A typical concentration-time profile qualitatively follows the well known HC/NO_x photochemical progression. The individual runs are available in a separate report (Arnts and Gay 1979). In Figure 2 the maximum O_3 concentration is plotted as a function of the initial C:NO_x (ppmC:ppm NO_x) for each of the compounds. The optimum O_3 production demonstrated by the olefins occurs at a C:NO_x ratio around 15:1, in fair agreement with that observed by Westberg (about 25:1) with terpinolene, d-limonene, and α-pinene (Coffey and Westberg 1977).

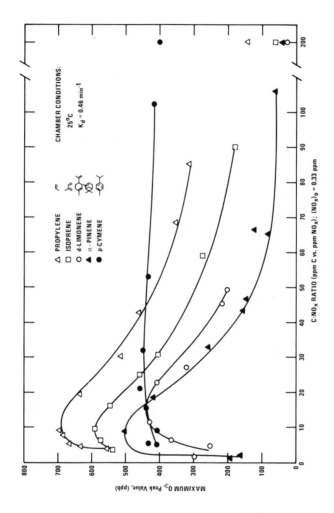

Figure 2. Effect of HC to NO_x ratio on O3 maximum.

Although the time to reach the O_3 maximum is reduced as the HC is increased, the concentration of O_3 formed after the optimum $C:NO_x$ ratio declines for the olefins. Furthermore, the O_3 decline is not constant relative to the other HC's, but seems to be proportional to the ozonolysis rate of the HC. This can be seen clearly by plotting the O_3 maximum for the runs made at $C:NO_x = 200$ (Figure 3). This figure demonstrates that the HC's with the highest ozonolysis rates are the most effective at suppressing O_3 buildup, especially when excess HC is present (high $C:NO_x$). Para-cymene on the other hand has a very slow reaction rate with O_3; thus, excess p-cymene at the high $C:NO_x$ ratios does not suppress O_3 buildup.

Another approach to demonstrating the efficiency of HC's in generating O_3 in a photochemical system is to ratio the maximum O_3 concentration to the concentration of HC reacted to that point. This is shown in Figure 4 where ppb O_3 produced:ppbC reacted is plotted as a function of the $C:NO_x$ ratio. At the lower $C:NO_x$ ratios, efficiencies of the HC's tend to merge to a relatively high value (0.2 ppb to 0.5 ppb O_3:ppbC); however at the high $C:NO_x$ ratios which are representative of rural and remote airsheds, efficiencies are diverging. The two monoterpenes, α-pinene and d-limonene, are one to two orders of magnitude less efficient respectively than propylene at a $C:NO_x$ of 200.

The products of irradiations of the biogenic HC's in the presence of NO_x are presented in Table I. On the whole, carbon balances were poor. For all of the terpenes, only 10% or less of the carbon reacted could be accounted for as products (slightly higher with myrcene). The products common to all the compounds included formic acid, formaldehyde, carbon monoxide, carbon dioxide, acetaldehyde, and PAN. Acetone was observed from the reaction of myrcene; methyl vinyl ketone and methacrolein were products of the isoprene photoxidation as supported by both IR and gas chromatography/mass spectrometry (GC/MS).

Where the balance of the reacted carbon is to be found is uncertain. The in-situ IR should reveal most gaseous products; however, only a few absorption lines remain unidentified. Even if small absorption coefficients are assumed for these lines, the bulk of carbon would still not be identified. This leads us to conclude that the undetected carbon has converted to aerosol or has deposited on the reactor walls. In either case the IR would not detect them.

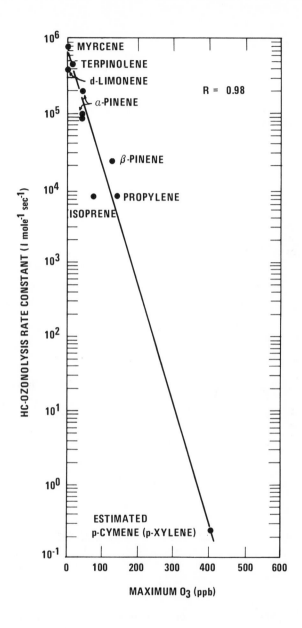

Figure 3.   HC ozonolysis rate vs. maximum O_3.
C:NO_x = 200; $(NO_x)_o$ = 0.33 ppm.

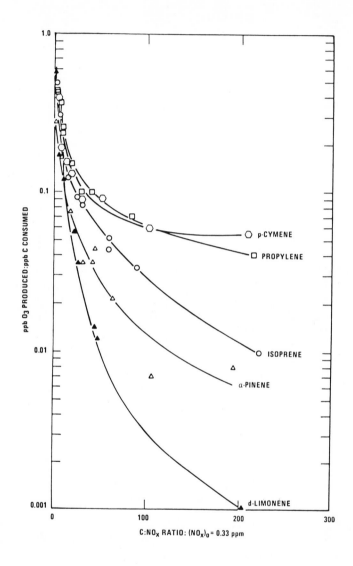

Figure 4. O_3 production efficiency relative to C consumption vs. $C:NO_x$.

TABLE I.   REACTIVITY AND PRODUCTS BY LONG-PATH IR AT 60 MIN

COMPOUND	C/NOx RATIO	%HC REACTED	% CARBON ACCOUNTED FOR
ISOPRENE 	31 10	68 100	31 44
MYRCENE 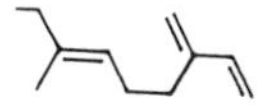	62	73	17
d-LIMONENE 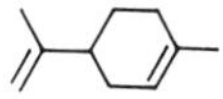	67	91	4
TERPINOLENE 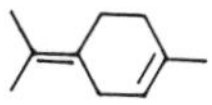	62	95	5
α-PINENE 	67	72	9
β-PINENE 	53	58	6
Δ³-CARENE 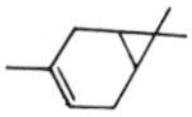	67	60	7

COMMON PRODUCTS:
HCHO
HCOOH
CO
CO$_2$
CH$_3$CHO
PAN

ACETONE (MYRCENE)

METHYL VINYL KETONE ⎫
METHACROLEIN ⎬ (ISOPRENE)

DISCUSSION

As stated previously, the results of the smog chamber
experiments cannot be directly extrapolated to the real
atmosphere. However, kinetic models have been developed to
simulate the chemistry observed in smog chamber studies. By
making some adjustments to the model, it can then be used to
predict real atmospheric conditions (Dodge 1977). One such
model is a 75-step reaction mechanism which describes the
atmospheric chemistry of an n-butane/propylene mixture (Dodge
1977). Although only a two HC mixture, the model accurately
simulates smog chamber experiments conducted on auto exhaust.

Presently a detailed mechanism for the photooxidation of isoprene or the terpenes cannot be written. Hence a model such as described above cannot be developed for the biogenic HC's. However, this study has established that the biogenic HC's certainly yield no more and probably much less O_3 than propylene. Since a model using propylene is available, propylene can be used as a surrogate for the biogenics. Such an application should allow us to place an upper limit on the O_3 potential of the biogenic HC's.

Based on detailed HC analyses of air in rural and remote areas not under the direct influence of urban centers, total nonmethane hydrocarbon (TNMHC) generally averages 40 ppbC to 100 ppbC. Of this amount, about 60% is paraffinic and 30% aromatic in character, and the balance olefinic (Arnts and Meeks 1980). Hence, isoprene and the monoterpenes generally account for less than 10% of the TNMHC.

Background nitrogen oxides ($NO + NO_2$) are perhaps equally difficult to define since they have both natural and anthropogenic sources. Although a large data base exists for these in urban areas, commercially available instruments are not sensitive enough to measure the low (less than 5 ppb) levels in rural areas. However, in a few intensive field studies where sensitive instrumentation has been used, the lowest NO_x levels east of the Mississippi River are on the order of 1 ppb to 2 ppb. However, even these levels seem to reflect the widespread combustion related sources of NO_x in the populated eastern half of the country. In contrast, levels in the sparsely populated West are sub-ppb in rural and remote areas. Recent measurements of NO_2 in the Colorado Rockies indicate clean continental air contains levels ranging from 15 ppt to 100 ppt (Noxon 1978).

For the purpose of examining a worst case (maximum O_3 generation), the Empirical Kinetic Modeling Approach (EKMA) photochemical model of Dodge was run with a 25% propylene/75% n-butane mixture to simulate a rural or clean atmosphere. Total nonmethane hydrocarbon levels were defined as 0 ppbC to 200 ppbC. Two ranges were used to encompass background levels of NO_x: 0 ppb to 0.14 ppb and 0 ppb to 2.1 ppb. The results are plotted in Figures 5 and 6. The isopleths show the maximum concentration of O_3 produced during one solar day given initial concentrations of NO_x and TNMHC. The model assumes no O_3 present initially, and uses 3% dilution/h of O_3 free air.

Figure 5 permits only 2 ppb of O_3 to be generated from our pristine air mass of 0.015 ppb to 0.1 ppb NO_x with 0.04 ppmC to 0.10 ppmC of TNMHC. Obviously, such an NO_x-deficient atmosphere will not contribute significantly to a tropospheric background of 25 ppb to 30 ppb of O_3.

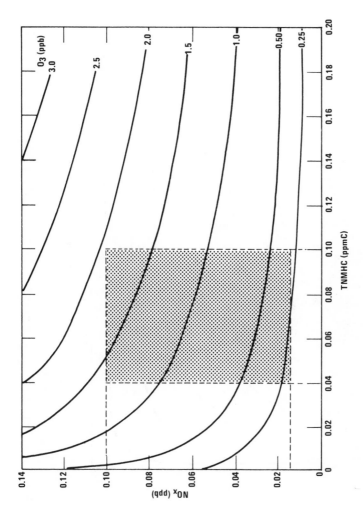

Figure 5. O₃ isopleths predicted from Dodge's photochemical model (75% n-butane/25% propylene).

127

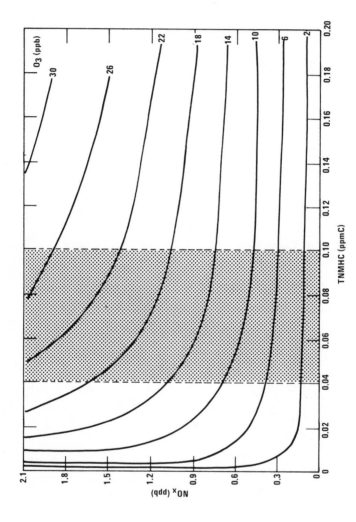

Figure 6. O₃ isopleths predicted from Dodge's photochemical model (75% n-butane/25% propylene).

128

If we examine the case of the 0 ppb to 2 ppb NO_x range at the same TNMHC range, significant concentrations of O_3 are generated. Between 0.04 ppmC to 0.10 ppmC about 20 ppb to 27 ppb of O_3 can be produced. As can be seen from these plots, O_3 generation is severely NO_x controlled and not a strong function of TNMHC.

Dodge has also run the above simulation at propylene/n-butane compositions from 1/99 to 75/25 to compare with the 25/75 mix described above. Interestingly, the concentrations of O_3 generated are olefin insensitive. Only slightly more O_3 was generated at the high olefin content and slightly less at the lower olefin percentages. Thus the presence of a biogenic olefin such as isoprene or α-pinene is of little consequence to the O_3 potential of an air mass that is composed of 60% paraffins and 30% aromatics.

Another approach to modeling the impact of biogenic HC emissions is to introduce the HC emissions with time instead of using ambient concentrations. Emissions from vegetation have been estimated (Arnts et al. 1977, Zimmerman 1979). For the purpose of assuming a worst case, a value of 60 $\mu g/m^2$-min was selected as representative of a high measured mass loading flux of HC. Again, to obtain maximum O_3 production, the NO_x flux was adjusted to yield a $C:NO_x$ ratio of 20:1. The NO_x flux was assumed to consist of 20% NO_2 and 80% NO. Propylene was again used in this simulation as in the previous model. A mixing height of 1.8 km (summertime afternoon average for North Carolina) was assumed. Ambient O_3 within the box was 40 ppb as was the dilution air which was exchanging at 5%/h. The photolysis of NO_2 and other photolytic species were varied through the simulations to follow that of a solar day.

The results of this simulation show that O_3 maximizes at about 70 ppb. Thus about 30 ppb of O_3 are generated photochemically against the 40 ppb background. Although this is not insignificant, isoprene and to an even lesser extent the terpenes will not produce this amount of O_3 (as was indicated by the experiments described herein). This, coupled with the assumptions of high HC flux and optimum $C:NO_x$ ratio, indicates the unlikely importance of biogenic HC's to rural O_3 formation.

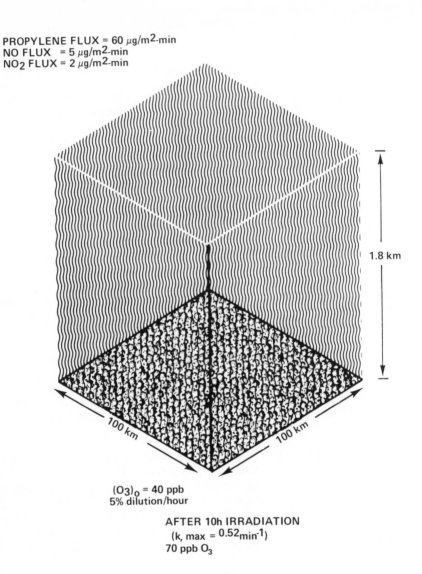

PROPYLENE FLUX = 60 $\mu g/m^2$-min
NO FLUX = 5 $\mu g/m^2$-min
NO_2 FLUX = 2 $\mu g/m^2$-min

1.8 km

100 km

100 km

$(O_3)_0$ = 40 ppb
5% dilution/hour

AFTER 10h IRRADIATION
(k, max = $0.52min^{-1}$)
70 ppb O_3

Figure 7.   Photochemical box model.

CONCLUSIONS

Laboratory experiments demonstrate that isoprene and monoterpenes are inefficient O_3 precursors relative to propylene in the HC/NO_x photochemical system. Their inefficiency increases with increasing carbon to NO_x ratio. The photochemical products of these compounds are still largely unknown.

Photochemical modeling, using a propylene/n-butane mixture as a surrogate biogenic HC, was performed to simulate the impact of these compounds on rural atmospheres. The first simulation demonstrated the maximum O_3 formed from an initial air mass containing 40 ppbC to 100 ppbC TNMHC and concentrations of NO_x ranging from 0.015 ppb to 2 ppb. These showed that the O_3 potential of such an air mass is severely NO_x dependent. Also, the model predicts that the O_3 potential is olefin insensitive. Thus the small measured ambient concentrations of olefinic natural HC's are not likely to have a significant effect on the O_3 potential of an air mass when compared with the large anthropogenically related paraffinic/aromatic fraction of the TNMHC.

The second simulation consisted of modeling the impact of a reported measured flux on O_3 production. An NO_x flux was assumed to yield a C:NO_x ratio of 20:1 to allow for optimum O_3 generation. An additional 30 ppb of O_3 was generated against a 40 ppb background. This is a surprisingly small amount considering the flux was assumed to be all propylene which is much more efficient than the biogenic HC's; also an unrealistic optimum C:NO_x ratio was used.

A more precise prediction of the impact of biogenic HC's on rural atmospheres will have to await a model based on the reaction mechanisms of these compounds. Presently, product distributions are not known, making a detailed mechanism impossible to write at this time. However, propylene as a surrogate biogenic HC can be used to predict upper limits to O_3 production.

REFERENCES

Arnts, R. R. and B. W. Gay, Jr. 1979. Photochemistry of Some Naturally Emitted Hydrocarbons. EPA-600/3-79-081. U.S. Environmental Protection Agency, Research Triangle Park, North Carolina.

Arnts, R. R., R. R. Seila, R. L. Kuntz, F. L. Mowry, K. R. Knoerr, and A. C. Dudgeon. 1977. Measurement of alpha-pinene fluxes from loblolly pine forest. Conference Proceedings, 4th Joint Conference on Sensing of Environmental Pollutants, New Orleans, November 6-11.

Arnts, R. R. and S. A. Meeks. 1980. Biogenic Hydrocarbon Contribution to the Ambient Air of Selected Areas. EPA-600/3-88-023. U.S. Environmental Protection Agency, Research Triangle Park, North Carolina.

Coffey, P. E. and H. Westberg. 1977. The issue of natural organic emissions. International Conference on Oxidants, 1976 - Analysis of Evidence and Viewpoints: Part IV. EPA-600/3-77-116. U.S. Environmental Protection Agency, Research Triangle Park, North Carolina.

Darnall, K. R., A. C. Lloyd, A. M. Winer, and J. N. Pitts, Jr. 1976. Reactivity scale for atmospheric hydrocarbons based on reaction with hydroxyl radical. Environ. Sci. and Technol. 7:692-696.

Dodge, M. C. 1977. Combined use of modeling techniques and smog chamber data to derive ozone-precursor relationships. International Conference on Photochemical Oxidant Pollution and Its Control, Preceedings 2:881-889. EPA-600/3-77-001b. U.S. Environmental Protection Agency, Research Triangle Park, North Carolina.

Glasson, W. A. and C. S. Tuesday. 1970. Hydrocarbon reactivities in the atmospheric photooxidation of nitric oxide. Environ. Sci. and Technol. 4:916-924.

Hanst, P. L., A. S. Lefohn, and B. W. Gay. 1973. Detection of atmospheric pollutants at parts-per-billion levels by infrared spectroscopy. Appl. Spectros. 22:188-198.

Lipeles, M., D. A. Landis, and G. M. Hidy. 1978. The formation of organic aerosols in a fast flow reactor. Adv. in Environ. Science. Volume 8. John Wiley, New York, New York.

Noxon, J. F. 1978. Tropospheric NO_2. J. Geophys. Res. 83:3051-3057.

Schuetzle, D. and R. A. Rasmussen. 1978. The molecular composition of secondary aerosol particles formed from terpenes. J. Air Pollution Control Assoc. 28:236-240.

Schwartz, W. 1974. Chemical Characterization of Model
Aerosols. EPA-650/3-74-011. U.S. Environmental Protection
Agency, Research Triangle Park, North Carolina.

Zimmerman, P. R. 1979. Determination of emission rates of
hydrocarbons from indigeneous species of vegetation in the
Tampa/St. Petersburg, Florida area. Final Report EPA
Contract No. 68-01-4432.

DISCUSSION OF PRESENTATION

[Author's note: In one or two instances, questions which
could not directly be addressed at the time of the symposium
are answered here for the reader.]

KAMENS: You used a 1.8 km mixing height in your box model.
Isn't that a bit high?

ARNTS: The value selected represents the afternoon summer
mixing height for central North Carolina.

KAMENS: Is that over the course of the 10-h irradiation?

ARNTS: Yes.

[Note: Since this symposium was held Dodge has performed a
more sophisticated simulation of that discussed here. She
started the simulation at a morning mixing height of 400 m
and increased it with time to the ultimate afternoon 1.7 km.
Although not directly comparable to the above simulation
since she used a higher HC flux of 100 $\mu g/m^2$-min,
nevertheless her simulation yielded only 8 ppb of O_3 more
than our simpler model.]

VOICE FROM AUDIENCE: What were the starting concentrations
of the O_3 precursors in the box, and what concentrations did
they ultimately achieve?

[Note: Following the symposium, an error was detected in the
box model calculations. The HC:NO_x ratio was actually 200:1
instead of 20:1. Hence the model was run again with the
corrected ratio and is presented in its correct form in
Figure 7. Therefore, in answer to the speaker's question,
the recalculated values are given: the propylene and NO_x
were both started at zero in the box model. At the time O_3
achieved its maximum concentration of 70 ppb some 12 h into
the simulation, propylene was present at 10.8 ppbC and NO_x
($NO + NO_2$) at 1.01 ppb.]

GRAEDEL: How does this situation apply to the one you presented earlier where you had the rural situation of less than 0.1 ppb NO_x?

ARNTS: The different methods described earlier were applied to demonstrate several effects. In the case of the O_3 isopleths [Figures 5 and 6], the simulations predict the maximum O_3 produced from a given <u>initial</u> mix of propylene/ n-butane and NO_x - a static system except for the dilution air exchanging with the system. No HC's and NO_x were added to the initial mixture. This approach was taken to simulate the <u>concentrations</u> of HC's and NO_x that are actually measured in clean continental air masses. Those runs clearly demonstrate the severe NO_x dependence of such an air mass. The box model uses measured <u>emission</u> <u>fluxes</u> of HC's. In order to demonstrate a worst case to generate O_3, an NO_x flux was calculated to yield an optimum C:NO_x ratio of 20:1. This ratio is in fact much too low for actual rural conditions (200:1 or higher would be more realistic but yield less O_3).

JEFFRIES: How would you characterize the reactivity of the biogenic HC's? Where would you place them?

ARNTS: If reactivity is defined as the rate of disappearance of the HC or the rate of NO to NO_2 conversion, then they are moderately to very reactive. In these respects isoprene is on a par with propylene. The monoterpenes are of higher reactivity.

JEFFRIES: How do you characterize them in terms of O_3 formation potential?

ARNTS: Relative to propylene, all of the compounds (except p-cymene) are less efficient in producing O_3. Their efficiency is strongly dependent on the C:NO_x ratio as shown in Figure 4.

OLLISON: Did you see aerosol formation in these experiments?

ARNTS: The limited volume of the irradiation mixes (250 1) did not allow us to collect aerosol samples.

OLLISON: You did not look for them?

ARNTS: No, aerosol measurements were not attempted due to the large volumes required. Aerosol formation was suspected in the long path IR photochemical cell since the HC reaction could not be accounted for by the formation of any new absorption bands. The products must therefore be removed from the gas phase either by aerosol formation and/or by deposition on the reactor walls.

WALKER: With relation to your chamber work, I appreciate the problems of working with low concentrations in chambers. But I would interpret the HC concentrations as having been fairly high in most of those runs plotted there.

ARNTS: Yes. They ranged from about 300 ppbC to 60 ppmC.

WALKER: But isn't it true for propylene or for any of them, that if you worked in a more dilute system you would get higher efficiency in terms of moles of O_3 per mole of carbon disappearing? Isn't this indicated by your Figure 4 where the efficiencies tend to converge at the lower HC concentrations?

BUFALINI: Yes, but that effect is not due to dilution; it is due to the carbon to NO_x ratio.

WALKER: It is. But your more dilute systems at the low $C:NO_x$ ratios are where the O_3 efficiencies tend to converge.

BUFALINI: We did do experiments at lower concentrations, and on a relative basis propylene was still much more photochemically reactive as far as O_3 production (Arnts and Gay 1979).

WALKER: But can you enhance the efficiency of all of them by dilution? Right down to zero?

BUFALINI: Well, the problem you encounter is contamination from the Teflon bags as was discussed this morning. At these lower HC concentrations, outgassing of contaminants can significantly affect O_3 generation. Some people have attempted to compensate for this effect by subtracting the blank. However, this is not entirely valid since there are synergistic effects occurring. So the big question remains as to how much O_3 is reactant generated and how much is bag artifact generated.

JEFFRIES: If you compare the reactivity of ethylene on a per carbon disappearance basis for the formation of O_3, it is more efficient than propylene. Ethylene reacts more slowly than propylene. Since ethylene takes a longer time to reach the same degree of loss, it has greater opportunity for secondary compounds - aldehydes - to contribute other radicals to the system to oxidize to produce the O_3. So the less reactive your initial reactant is, the more efficient it appears on a per disappearance basis in terms of forming O_3.

VOICE FROM AUDIENCE: And by turn, then, butane is more efficient than ethylene.

JEFFRIES: Yes.

BUFALINI: The question of chamber contamination effects still needs to be answered. We have noted and more recently Harvey Jeffries has also observed NO_x regeneration. During the course of irradiations, it appears that NO_x is lost to the walls as nitric acid (HNO_3) and later comes back off as NO_2. Thus you get a considerable amount of O_3 generated in these systems. One approach to eliminating wall effects may be to emit known quantities of pollutants and an inert tracer to the atmosphere and follow the chemistry of the air mass.

OLLISON: One other question on the concentration of the terpenes: if you use relatively high concentrations does that enhance aerosol formation relative to lower or natural concentrations?

ARNTS: That should be dictated in large part by the vapor pressure of the products. If at the lower concentrations of terpenes the products generated are above their vapor pressures, then they should generate aerosols in a manner similar to the high concentration experiments. If they are below their vapor pressures, then those compounds should remain in the gas phase and probably continue to participate in the photochemical chains.

ZIMMERMAN: If you had to make an estimate of how many molecules of O_3 you could get from one molecule of isoprene what would you estimate from these experiments?

ARNTS: Isoprene, as are the other HC's, was dependent on the $C:NO_x$ ratio with respect to O_3 production efficiency. It ranged from a high of about 0.5 ppb O_3 produced at the O_3 maximum for every ppbC consumed to 0.01 ppb O_3:ppbC at $C:NO_x$ ratios from 4 to 200, respectively.

BUFALINI: We are aware of the numbers you have been using [to P. Zimmerman]. How did you arrive at them?

ZIMMERMAN: The numbers were developed from photochemical models used at the National Center for Atmospheric Research (NCAR). The modelers assume an ambient concentration of 10 ppt NO and complete photochemical oxidation of the isoprene to CO.

JEFFRIES: That only works if you keep nitrogen (NO) in the system.

LUDLUM: That's right. But that's exactly the type of thing Joe [Bufalini] was pointing out in the paper he published: he said 1 ppm of $C_{10}H_{16}$ will give you 13 ppm O_3. And so if

136

you had 1 ppm of isoprene, or, say terpene, you have that much greater O_3 forming potential.

BUFALINI: The purpose of that paper was to put upper limits on the amount of O_3 that could be formed. In practice these limits are never really achieved, chiefly because there is not enough NO_x present.

[Editor's Note: The paper referenced by Ludlum was by J. J. Bufalini, T. A. Walter, and M. M. Bufalini. 1976. Environ. Sci. Technol. 10:908. The actual number of O_3 produced by a $C_{10}H_{16}$ olefin would be nC + nH −1 (for addition of OH) or 25].

16. AEROSOLS AND CARBON BALANCE IN THE PINENE-NO$_x$ PHOTOCHEMICAL SYSTEM

Murray J. Kaiserman and Eric W. Corse. Northrop Services, Inc., P.O. Box 12313, Research Triangle Park, North Carolina 27709

ABSTRACT

Alpha-pinene-NO$_x$ and beta-pinene-NO$_x$ mixtures were irradiated in a large smog chamber. Relative humidity changes did not affect the system. The results indicated that aerosol formation was not important to the carbon balance, with 2% to 3% of the carbon initially present being converted to aerosols. Carbon monoxide was determined to be the ultimate fate of pinene carbon with a conversion rate of 1% carbon/h. Only 20% to 40% of the carbon initially present could be accounted for.

INTRODUCTION

The photochemical reaction of natural emissions has been shown to account for the blue haze found in nature (Lillian 1972, Went 1960, Ripperton et al. 1971, Wilson et al. 1972). However, a number of questions concerning both the effects and the fate of natural emissions in the atmosphere have been raised (Lillian 1972, Went 1960, Ripperton et al. 1971, Wilson et al. 1972, Maugh 1975). Uncertainty exists as to whether the naturally occurring photooxidation of terpenes serves as a source or as a sink for ozone (O$_3$) (Stasiuk and Coffey 1974, Cronn et al. 1977, Lillian 1972, Lipeles et al. 1978, Gay and Arnts 1977). In addition, the role of natural emissions in the conversion of gas to particles and the role of oxygenated intermediates in the formation of aerosols remain to be established (Maugh 1975, Japar et al. 1974).

Concerning the fate of natural emissions, the major question
is whether or not carbon monoxide (CO) is the ultimate fate
of most of the naturally emitted carbon, instead of aerosols
as has previously been believed (Lillian 1972, Went 1960,
Wilson et al. 1972, O'Brien et al. 1975, Cronn et al. 1977,
Gay and Arnts 1977, Zimmerman et al. 1978, Cox et al. 1980,
Goetz and Pueschel 1967, Groblicki and Nebel 1971).

A considerable amount of data relative to these questions
has been collected, but no systematic chamber study has been
conducted which would determine if either aerosol or CO was
the ultimate fate of the emitted terpene (Lillian 1972, Went
1960, Ripperton et al. 1971, Wilson et al. 1972, Maugh 1975,
Stasiuk and Coffey 1974, O'Brien et al. 1975, Cronn et al.
1977, Lipeles et al. 1978, Gay and Arnts 1977, Japar et al.
1974, Zimmerman et al. 1978, Cox et al. 1980, Goetz and
Pueschel 1967, Groblicki and Nebel 1971). Thus, this study
was undertaken to examine the irradiation of mixtures of
either α- or β-pinene and nitrogen oxides (NO_x) in a smog
chamber. In addition to monitoring both CO and aerosols,
testing was conducted to study the effect of water on the
reactivity of the pinenes.

EXPERIMENTAL PROCEDURES

The smog chamber is an aluminum and Teflon box, 8 ft x 8
ft x 6.25 ft high (2.4 m x 2.4 m x 1.9 m), with an effective
volume of 400 ft^3 (11.3 m^3). Teflon windows mounted against
the aluminum frame permit the irradiation of the chamber
interior. Two light banks on each side of the chamber,
consisting of 56 black lights (F96T8·BL, General Electric)
and 9 sun lamps (FS 20, Westinghouse) yield a volume
integrated value for k_1 of 0.4 min^{-1}. The chamber
characteristics are checked by periodically irradiating
either "clean air" or standard propylene–NO_x mixtures.

Thermocouple probes connected to a digital volt meter
(DVM) measure the temperature at various points within the
chamber. A CTE dew point hygrometer measures the relative
humidity within the chamber. Mounted inside the chamber and
sealed to an access port is an optical particle counter (OPC)
(Climet Model 208A) sealed in a Tedlar bag. An electrical
aerosol analyzer (EAA) (Thermo Systems, Inc.) samples
aerosols through a bag fitting. The remainder of the instruments are connected to the chamber by Teflon sampling lines.

The EAA, CTE, and DVM are driven by a Digital PDP 11/40
computer system which also accepts and stores the signals.
In addition, the computer drives a multichannel analyzer
(MCA) to accept pulses from the OPC.

A flame ionization detector gas chromatograph (FID-GC) (Perkin Elmer Model 900) monitors the terpenes. The chromatograph is equipped with a stainless steel column packed with 3.8% SE30 silicone oil on 80 to 100 mesh Chromosorb W. Carbon monoxide, methane (CH_4), and total nonmethane hydrocarbon (THC) are monitored by a Beckman Model 6800 air quality chromatograph. Ozone is monitored by a Bendix Model 8002 O_3 Monitor. Nitric oxide (NO) and NO_x are monitored by two Bendix $NO-NO_2-NO_x$ analyzers. These instruments are either checked or calibrated before each irradiation.

Aerosol samples are collected at the end of the irradiation by passing the chamber contents through a Millipore Corporation Type LS Mitex Teflon filter with a 47 mm diameter and a 5 µm pore size.

Aldehydes are analyzed by the 3-methyl-3-benzothiazolone hydrazone (MBTH) method (Sawicki et al. 1961). Nitrogen dioxide is analyzed by the Saltzman method (Saltzman 1954).

Both α-pinene, supplied by Aldrich Chemical Co., Inc., and β-pinene, supplied by SCM Organic Chemicals, were analyzed by gas chromatographic techniques and found to be better than 98% pure. Nitric oxide was supplied in cylinders by Scott Specialty Gases.

RESULTS

Data for the effect of relative humidity on the reactivity of the tepenes are presented in Tables I and II. For α-pinene, a change in relative humidity was found to affect neither the terpene half-life nor the amount of O_3 formed. For β-pinene, an increase in the relative humidity resulted in a decrease in the terpene half-life. However, no change was observed in the amount of O_3 formed. Since the reactivity of both pinenes seemed insensitive to changes in relative humidity, the study was continued under conditions of high relative humidity (>75%).

At constant pinene concentrations, an increase in the NO_x concentration resulted in a decrease in the terpene half-life. The amount of O_3 formed also increased. These data are presented in Tables III and IV.

Data for the effect of varying the pinene concentration at constant NO_x concentration (either 0.1 or 0.2 ppm) are presented in Tables V and VI. A decrease in the pinene concentration decreased pinene half-life; however, the amount of O_3 formed increased.

Since the concentration profiles of both the reactants and products exhibited similar trends regardless of initial concentrations, only two representative examples of the data were chosen for presentation. The variations of the pinene, NO, and O_3 concentrations vs. time are presented in Figures 1 and 2. The O_3 concentration reached a maximum value during the first 20 to 30 min. The NO concentration quickly fell to less than 0.001 ppm, and the pinene concentration gradually decreased to less than 0.10 ppm.

Carbon monoxide and ozone concentration vs. time profiles are presented in Figures 3 and 4. The concentration of CO increased in a linear manner once the O_3 maximum concentration was reached. Experiments were also performed to determine if CO was indeed a photolysis product. If the lights were turned off, no formation of CO occurred, even overnight. Once the lights were turned on, CO formation resumed.

Data obtained by the EAA are shown in Figures 5 and 6. The majority of particles formed (⩾99%) had a diameter of less than 1.0 µm. The largest fraction of these particles had a range of diameters from 0.133 µm to 0.562 µm. In addition, the total volume of these particles per cc remained constant during the first 4 h to 6 h after the maximum value of total volume per cc was reached.

Tables VII and VIII present data for a carbon balance. The mass of the collected aerosols were found to be in good agreement with the EAA data, but only accounted for 2% to 3% of the total carbon. In addition, only 20% to 40% of the carbon initially present could be accounted for.

TABLE I. THE EFFECT OF RELATIVE HUMIDITY ON IRRADIATIONS
OF MIXTURES OF ALPHA-PINENE AND NO_x

Date	Initial Pinene	Initial NO_x	Time @ 1/2 Terpene	Relative Humidity	Maximum O_3	Pinene/NO_x
	(ppm)	(ppm)	(min)	(%)	(ppm)	(ppm/ppm)
2-28-79	2.14	0.338	65	39.8	0.130	6.3
2-21-79	2.05	0.340	64	43.9	0.140	6.0
3-14-79	2.61	0.335	54	81.4	0.125	7.3
3-05-79	2.19	0.345	66	82.7	0.119	6.4

TABLE II. THE EFFECT OF RELATIVE HUMIDITY ON IRRADIATIONS
OF MIXTURES OF BETA-PINENE AND NO_x

Date	Initial Pinene (ppm)	Initial NO_x (ppm)	Time @ 1/2 Terpene (min)	Relative Humidity (%)	Maximum O_3 (ppm)	Pinene/NO_x (ppm/ppm)
2-06-79	1.78	0.308	134-137	12.7	0.192	5.8
4-23-79	1.77	0.332	121	45	0.230	5.5
2-15-79	2.07	0.314	133	38.4	0.214	6.6
3-22-79	2.17	0.333	92	91.2	0.170	6.52
3-19-79	2.08	0.347	90	79.1	0.239	6.0

TABLE III. ALPHA-PINENE AND NO_x MIXTURES
AT CONSTANT ALPHA-PINENE

Date	Initial Pinene (ppm)	Initial NO_x (ppm)	Time @ 1/2 Terpene (min)	Relative Humidity (%)	Maximum O_3 (ppm)	Pinene/NO_x (ppm/ppm)
3-26-79	2.46	–	200	84	–	–
4-12-79	2.44	–	208	78.9	–	–
5-24-79	2.05	0.106	104	85.2	0.045	19.3
4-30-79	2.36	0.106	188	81.8	0.033	22.3
5-29-79	2.76	0.220	92	86.4	0.080	12.5
5-03-79	2.12	0.206	90	77	0.085	10.3
3-05-79	2.19	0.345	66	82.7	0.119	6.4
3-14-79	2.61	0.335	54	81.4	0.125	7.8

TABLE IV. BETA-PINENE-NO$_x$ MIXTURES AT CONSTANT BETA-PINENE

Date	Initial Pinene (ppm)	Initial NO$_x$ (ppm)	Time @ 1/2 Terpene (min)	Relative Humidity (%)	Maximum O$_3$ (ppm)	Pinene/NO$_x$ (ppm/ppm)
3-29-79	1.77	–	260	80.9	–	–
4-02-79	1.83	–	261	84.9	–	–
5-21-79	1.31	0.107	154	86.9	0.119	12.2
5-10-79	1.31	0.108	–	78.4	0.109	12.1
5-14-79	1.70	0.207	128	78.2	0.170	8.2
5-07-79	3.54	0.203	115	79	0.170	17.4
3-19-79	2.08	0.347	90	79.1	0.239	6.0
3-22-79	2.17	0.333	92	91.2	0.170	6.5

TABLE V. ALPHA-PINENE-NO$_x$ MIXTURES AT CONSTANT NO$_x$

Date	Initial Pinene (ppm)	Initial NO$_x$ (ppm)	Time @ 1/2 Terpene (min)	Relative Humidity (%)	Maximum O$_3$ (ppm)	Pinene/NO$_x$ (ppm/ppm)
7-09-79	0.749	0.107	–	83.9	0.047	7.0
7-16-79	1.06	0.103	84	95.5	0.073	10.3
7-03-79	1.0	0.105	81	84.1	0.057	9.5
5-24-79	2.05	0.106	104	85.2	0.045	19.3
4-30-79	2.36	0.106	188	81.8	0.033	22.3
5-03-79	2.12	0.206	90	77	0.085	10.3
5-29-79	2.76	0.220	92	86.4	0.080	12.5
6-07-79	1.67	0.240	36	81.6	0.128	7.0
6-26-79	1.30	0.208	42	93.5	0.109	6.3
6-11-79	1.13	0.195	45	81.6	0.115	5.8

TABLE VI. BETA-PINENE-NO$_x$ MIXTURES AT CONSTANT NO$_x$

Date	Initial Pinene (ppm)	Initial NO$_x$ (ppm)	Time @ 1/2 Terpene (min)	Relative Humidity (%)	Maximum O$_3$ (ppm)	Pinene/NO$_x$ (ppm/ppm)
4-21-79	0.971	0.200	77	91.1	0.189	4.9
6-14-79	0.859	0.200	–	83.1	0.227	4.3
5-14-79	1.70	0.207	128	78.2	0.170	8.2
5-07-79	3.54	0.203	115	79	0.170	17.4
5-10-79	1.31	0.108	–	78.4	0.109	12.1
5-21-79	1.31	0.107	154	86.9	0.119	12.1
6-18-79	0.929	0.103	100	81.2	0.152	9.0
7-06-79	0.729	0.105	104	84.7	0.175	6.9

TABLE VII. CARBON BALANCE FOR THE IRRADIATION OF ALPHA-PINENE IN THE PRESENCE OF NO$_x$ (MAY 3, 1979)

Carbon[1] As Aldehyde (ppm)	Carbon[2] As CO (ppm)	Carbon[2] As THC (ppm)	Carbon[3] As Aerosol (ppm)	Total Carbon (ppm)	Total Carbon Irradiated (ppm)	Total Carbon Accounted For (%)
0.323	1.087	2.360	0.630	4.400	21.2	20.7

[1] MBTH method
[2] Beckman 6800 Air Quality Chromatograph
[3] Collected at the end of the irradiation

TABLE VIII. CARBON BALANCE FOR THE IRRADIATION OF
BETA-PINENE IN THE PRESENCE OF NO_x (MAY 10, 1979)

Carbon[1] As Aldehyde (ppm)	Carbon[2] As CO_2 (ppm)	Carbon[2] As NMH (ppm)	Carbon[3] As Aerosol (ppm)	Total Carbon (ppm)	Total Carbon Irradiated (ppm)	Total Carbon Accounted For (%)
0.342	0.751	2.70	0.360	4.153	13.1	31.7

[1]MBTH method
[2]Beckman 6800 Air Quality Chromatograph
[3]Collected at the end of the irradiation

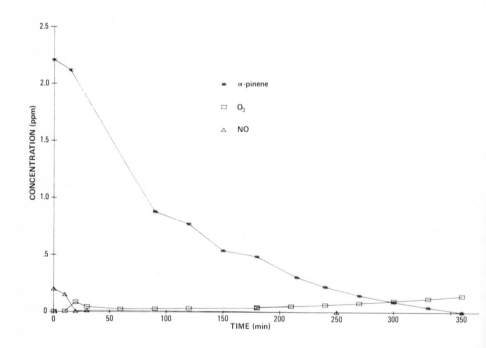

Figure 1. The variation of α-pinene, NO, and O_3
versus time for May 3, 1978.

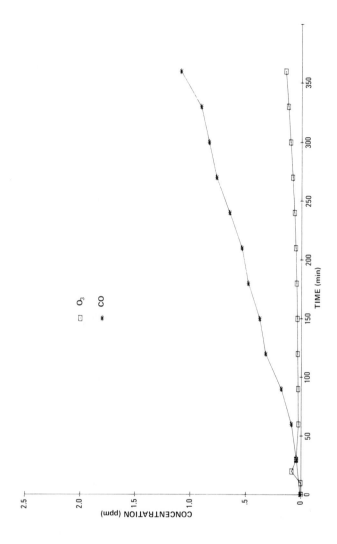

Figure 2. The variation of O_3 and CO versus time for May 3, 1979.

147

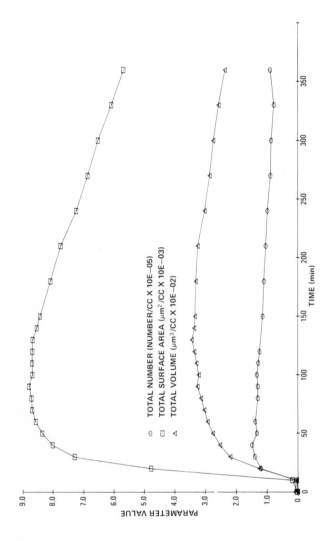

Figure 3. The variation of total number, total surface area, and total volume versus time for May 3, 1979.

148

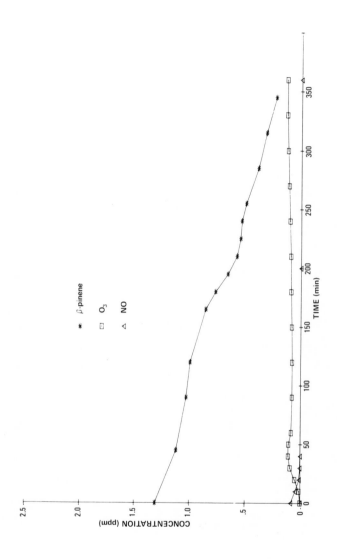

Figure 4. The variation of β-pinene, NO, and O₃ versus time for May 10, 1979.

149

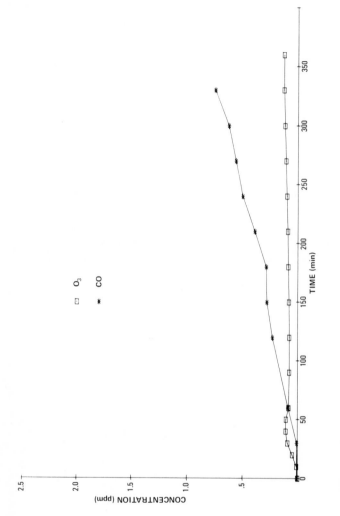

Figure 5. The variation of O_3 and CO versus time for May 10, 1979.

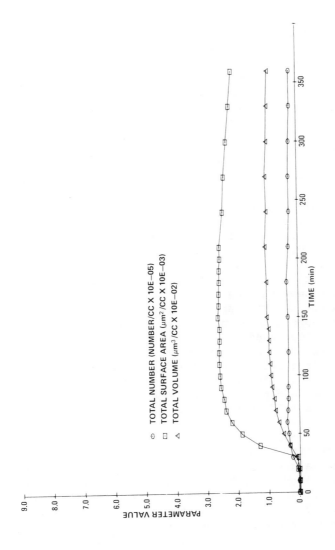

Figure 6. The variation of total number, total surface area, and total volume versus time for May 10, 1979.

151

CONCLUSIONS

Earlier work in this area had shown that the amount of O_3 formed during irradiations of pinene–NO_x mixtures depended on the carbon to NO_x ratio (Gay and Arnts 1977, Arnts and Gay 1979). The decision to irradiate pinene–NO_x mixtures with carbon to NO_x ratios of at least 50 to 1 was made because these mixtures would represent minimum ambient conditions (Arnts and Gay 1979). In addition, O_3 formation data could be used as a check of the chamber properties. The results of this study confirmed the opinion that at ambient concentrations, the pinenes act as a sink rather than as a source for O_3 (Gay and Arnts 1977, Arnts and Gay 1979). The results also show that changes in the relative humidity had little effect on the amount of O_3 formed during the irradiation of pinene–NO_x mixtures. However, the β–pinene half–life decreased with increased relative humidity. At present, this effect is unexplained.

The results further indicate that CO, not aerosol formation, is the ultimate fate of the pinenes. This conclusion disagrees with most of the earlier work (Lillian 1972, Went 1960, Wilson et al. 1972, O'Brien et al. 1975, Cronn et al. 1977, Gay and Arnts 1977, Goetz and Pueschel 1967, Groblicki and Nebel 1971). Both mass and EAA measurements show that only 2% to 3% of the carbon initially present in the chamber can be accounted for. Furthermore, this value represents an upper limit since the aerosols were assumed to be composed of only carbon. Although unexpected, these results are not surprising since this study is the first to monitor both directly and continuously aerosol formation without significantly affecting the system. Secondly, the data on O_3 formation agrees with previous work (Gay and Arnts 1977, Arnts and Gay 1979).

It is possible to convert both mass and EAA data to either visibility measurements or light scattering coefficient values (b_{scat}) (Charlson et al. 1967). For our representative α– and β–pinene systems, the visibility is reduced to 3.5 mi and 6 mi respectively. Particles of the size range produced in the chamber can produce a blue haze. It is noteworthy that a gas to particle conversion of only 2% to 3% can reduce visibility to these levels. The future addition of a nephelometer to our monitoring system will enable us to measure directly b_{scat} values.

The results obtained also confirmed the assertion that the ultimate fate of the pinenes was CO (Zimmerman et al. 1978). A photolytic conversion rate for either α– or β–pinene C to CO of about 1% C/h was observed. This conversion rate was independent of relative humidity. It was also found that the lights must be turned on for CO formation to occur. If the lights were turned off, the CO concentration remained constant even overnight. When the lights were turned on in the morning, CO began to be produced.

A higher conversion of pinene to CO would have been useful, but the dilution of the chamber contents would have been so high that the results would have been meaningless. Instead, it can only be surmised that CO precursors persist overnight, and that these precursors are not aerosols.

Finally, the poor carbon balance must be considered. Since it was possible to monitor aerosols, it cannot be assumed that any unaccounted carbon was converted to aerosols and deposited on the chamber walls (Gay and Arnts 1977). Only 20% to 40% of the carbon initially present in the chamber could be accounted for. About 10 percent of the carbon is reported as total hydrocarbon (THC), which includes any residual pinene, oxygenated hydrocarbons, nitrogenated hydrocarbons, and all other hydrocarbons present within the chamber. Determining exactly what these species are is important because they have different responses in a flame ionization detector. Knowledge of both the species and their concentrations will improve the carbon balance.

In summary, the following conclusions can be made:

- At ambient concentrations, the pinenes are a sink for O_3.

- Aerosols are not important to the carbon balance.

- The ultimate fate of the pinenes is CO, but the conversion to CO occurs at a minimum rate of about 1% C/h of irradiation.

- Work must be done to determine the types and concentrations of the oxygenated species that are produced.

REFERENCES

Arnts, R. R. and B. W. Gay. 1979. Photochemistry of Some Naturally Emitted Hydrocarbons. EPA-600/3-79-081. U. S. Environmental Protection Agency, Research Triangle Park, North Carolina.

Charlson, R. J., H. Horvath, and R. F. Pueschel. 1967. The direct measurement of atmospheric light scattering coefficient for studies of visibility and pollution. Atmos. Environ. 1:469-478.

Cox, R. A., R. G. Derwent, and M. R. Williams. 1980. Atmospheric photooxidation reactions. Rates, reactivity, and mechanism for reaction of organic compounds with hydroxyl radicals. Environ. Sci. Technol. 14:57-61.

Cronn, D. R., R. J. Charlson, R. L. Knights, A. L. Crittenden, and B. R. Appel. 1977. A survey of the molecular nature of primary and secondary components of particles in urban air by high-resolution mass spectrometry. Atmos. Environ. 11:929-937.

Gay, B. W. and R. R. Arnts. 1977. The chemistry of naturally emitted hydrocarbons. In: International Conference on Photochemical Oxidant Pollution and Its Control, Proceedings 2:745-751. EPA-600/3-77-001b. U.S. Environmental Protection Agency, Research Triangle Park, North Carolina.

Goetz, A. and R. Pueschel. 1967. Basic mechanisms of photochemical aerosol formation. Atmos. Environ. 1:287-306.

Groblicki, P. J. and G. J. Nebel. 1971. The photochemical formation of aerosols in urban atmospheres. In: Chemical Reactions in Urban Atmospheres, C. S. Tuesday, ed. Elsevier, New York, New York. pp. 241-267.

Japar, S. M., C. H. Wu, and H. Niki. 1974. Rate constants for the gas phase reaction of ozone with α-pinene and terpinolene. Environ. Lett. 7(3):245-249.

Lillian, D. 1972. Formation and destruction of ozone in a simulated natural system (Nitrogen dioxide + α-pinene + hν). In: Photochemical Smog and Ozone Reactions, R. F. Gould, ed. ACS Advances in Chemistry Series No. 113, pp. 211-218.

Lipeles, M., D. A. Landis, and G. M. Hidy. 1978. The Formation of Organic Aerosols in a Fast Flow Reactor. Advances in Environmental Sciences. Vol. 8. John Wiley, New York, New York.

Maugh, T. H., II. 1975. Air pollution: Where do hydrocarbons come from? Science. 189:277-278.

O'Brien, R. J., J. R. Holmes, and A. H. Bockian. 1975. Formation of photochemical aerosol from hydrocarbons:

Chemical reactivity and products. Environ. Sci. Technol. 9:568-576.

Ripperton, L. A., H. Jeffries, and J. J. B. Worth. 1971. Natural synthesis of ozone in the troposphere. Environ. Sci. Technol. 5:246-248.

Saltzman, B. E. 1954. Colorimetric microdetermination of nitrogen dioxide in the atmosphere. Anal. Chem. 26:1949-1955.

Sawicki, E., T. R. Hauser, T. W. Stanley, and W. Elbert. 1961. The 3-Methyl-3-benzothiazolone hydrazone test. Anal. Chem. 33:93-96.

Stasiuk, W. N., Jr. and P. E. Coffey. 1974. Rural and urban ozone relationships in New York State. J. Air Pollution Control Assoc. 24:564-568.

Went, F. W. 1960. Blue hazes in the atmosphere. Nature. 187:641-643.

Wilson, W. E., W. E. Schwartz, and G. W. Kinzer. 1972. Haze Formation - Its Nature and Origin. Report to U.S. Environmental Protection Agency. CPA 70 - Neg 172 and CRC-CAPA 6-68-3 (NTIS 212609).

Zimmerman, P. R., R. B. Chatfield, J. Fishman, P. J. Crutzen, and P. L. Hanst. 1978. Estimates on the production of CO and H_2 from the oxidation of hydrocarbon emissions from vegetation. Geophy. Res. Lett. 5(8):679-682.

DISCUSSION OF PRESENTATION

BUFALINI: As chairman, I'll ask the first question. I really don't think that's an adequate test for the Zimmerman model, because the aerosols have no place to hide. The material has no place to hide in a smog chamber.

I think a more appropriate test would be to put in aerosols, that is, perhaps other types of aerosols, which would provide a hiding place for the organic matter. Because, if you take any hydrocarbon and irradiate it in a smog chamber, I think the chemistry would suggest that it's going to have to go to CO and CO_2 and H_2O. It's got no place else to go.

Anybody want to argue that point?

LUDLUM: I want to clarify what you're saying. Are you saying that you need another aerosol in there, as kind of a nucleus for it to form on?

BUFALINI: Well, I offer that only as a suggestion. I think that certainly a smog chamber is not like ground, or oceans, or trees; so, in the atmosphere, these aerosols do have a place to deposit. And once they're deposited, it's doubtful that they'll come back up into the gas phase.

In this particular case, they have no place to hide; even though they deposit, they're going to have to come back out, or oxidize on the surface. But they will oxidize. I don't think that they're going to just stay there on the surface.

OLLISON: Are you saying that the smog chamber has a surface-to-volume ratio comparable to the atmosphere?

BUFALINI: No.

OLLISON: You say, in nature, in the atmosphere, the aerosols are going to the surface much more quickly than they do in a smog chamber? Is that why you need this added sulfate aerosol or whatever?

LUDLUM: He maintains that they're removing it, so they can push the reaction. He's saying you have a place to shift the equilibrium.

KAISERMAN: In a smog chamber, the oxygenated compounds would probably have to go to the wall. There's some debate on that; but we will probably have to do an analysis of our chamber contents.

DIMITRIADES: You mentioned a few numbers. Could you summarize for me what proportion of the initial reaction would be converted into aerosol, and what proportion into CO? As a generalization.

KAISERMAN: Say, about 1 percent per hour, for CO; so, in 6 hours, 6 percent. That would probably be a reasonable estimate.

For aerosol, I'd say I could account for about 2% total (over the course of the irradiation).

VOICE FROM AUDIENCE: That's constantly being illuminated in the smog chamber?

KAISERMAN: Yes, it's constantly illuminated.

156

DIMITRIADES: Now, that's a really significant amount of aerosol. But again, the data that we have on organic aerosols in the remote areas don't support that much aerosol; organic aerosol.

KAISERMAN: I don't think you can make this kind of comparison.

LUDLUM: How are you comparing the two? What comparison are you using?

W. E. WILSON: What was the 30%?

DIMITRIADES: No, not 30%; 2% per hour.

KAISERMAN: No, this is 2% total. It looks like it's constant over the course of the irradiation.

LUDLUM: Total; that's what he said.

KAISERMAN: Yes.

DIMITRIADES: Oh. I see.

KAISERMAN: For the EAA data--once it maximizes, once the total surface area maximizes, once the total volume maximizes--it's essentially constant until it starts settling out.

DIMITRIADES: Oh, I see. I see. You don't have continuous--

KAISERMAN: No. It may be a dynamic process and it may be settling out.

ALTSHULLER: Just to clarify this point, assume that yields from these chamber runs could be extracted from the atmosphere. You're starting with 100 $\mu g/m^3$ of terpenes; 2% of it, 2 μg to be yielded, become aerosol, etc.

KAISERMAN: Yes, assuming nothing happens.

BUFALINI: But, even if you accept the fact that all these data are correct, you still have to explain the Gay-Arnts data. Obviously there is a problem, because their material disappears. They claim that it went to aerosol. You cannot ascribe the material to being in the gas phase, even if you're extremely liberal with molar absorption coefficients.

So, where did it go?

[Author's Note: Gay and Arnts assume that since the missing material is not in the gas phase, it must be deposited on the chamber walls. They never confirmed this assumption.]

ALTSHULLER: Well, it depends obviously on what the other products are. But let's assume that the other products were highly polar material. What's so mysterious about them attaching to the walls and staying attached to the walls? We did experiments 15 years ago on glass, where we got formaldehyde in glass, and could never get it off again.

So what's so wonderful about aluminum in terms of being a reversible substrate?

W. E. WILSON: You've got a substantial amount of your carbon that's unaccounted for, for instance, seen in the infrared. It doesn't show up in anything that you can measure in the aerosol, in your work.

But, from the size of the aerosol and what we know about the products of ozone-pinene reactions, these organic materials have a fairly high vapor pressure. That is why you don't have a lot of small particles being formed. You have to wait until you get big particles, as a function of the vapor pressure.

That means you're going to have a lot of the material going to the wall. The smaller the smog chamber, the more vigorous the stirring that comes from thermal gradients (I presume you've got your stirring fan turned off, but you get stirring from the thermal gradients) the more material lost on the wall.

In the early studies of the effect of stirring we saw that for pure organic aerosols (e.g., toluene), tremendous amounts were lost to the walls by stirring.

In the early studies of aerosol potential of organic compounds, we used a small amount of SO_2 to give us a sulfuric acid (H_2SO_4) nucleus for the stuff to form on, because it always is that way.

So I would expect, you've got a lot of material lost on the wall. You might try a small amount of SO_2, so you have a H_2SO_4 nucleus, and see if it improves your organic aerosol recovery.

ALTSHULLER: If we're going to say we've got only 20% or 30% accounted for; let's take a wild assumption: all the rest of it is on the wall, okay? Is it possible, physically speaking, simply to put a shallow trough into the bottom of this chamber? This trough would represent an appreciable amount of the total surface area. Would you then be able to wash out that trough, and get something analytically from it?

With a glass flask you can obviously smoke the flask. But what you need to do now is get some sort of a surface you can remove material from.

KAISERMAN: We have discussed this with our GC/MS expert, and he said, the whole wall. Essentially, I think we'd have to clean up the whole surface area.

What eventually we will have to do is to go in and wash down the walls.

VOICE FROM AUDIENCE: Well, do it in a small chamber.

KAISERMAN: I think that's been done.

BUFALINI: Yes, the problem you run into with a small chamber is that you're running rather high concentrations on a part-per-billion carbon. And of course, you always run into the same problem, as someone else pointed out earlier, of whether or not the chemistry is going to be the same at 20 ppm carbon, compared to 20 ppb carbon. Certainly, the product distribution seems to change to some extent.

The advantage of doing it in a large chamber is you have all this volume; when working with a smaller chamber, you have to start with very high concentrations in order to get the analytical techniques. However, I think if we listen to the next paper, which Les Hull is going to be giving, he may clarify at least part of the problems.

HULL: Let me make one comment here. The problem with going to a smaller chamber is that the aerosol formation is going to be concentration dependent. As I will talk about in a little while, we want to make estimates of the vapor pressures of these compounds. Unfortunately, the chamber studies tend to be done in the range where these compounds are at about their saturation vapor pressure. So if you try to get realistic concentrations in the part per billion range, you have large chambers and it's difficult to rinse the walls. If you go to small chambers where you can rinse the walls, which I essentially have done, then you're going to get a lot of aerosol, which may not reproduce the same kind of chemistry as occurs under the low concentration system.

17. TERPENE OZONOLYSIS PRODUCTS

Leslie A. Hull. Department of Chemistry, Union College,
Schenectady, New York 12308

ABSTRACT

This paper presents the results of a product study of the
gas phase ozonolysis of alpha-pinene and beta-pinene with
reactant concentrations above 100 ppm. The reactions show a
0.9:1 to 1.4:1 consumption of terpene:ozone. With filtration
of the product mixtures, most of the mass of the terpene
consumed can be accounted for as collected product. Analysis
of the product mixtures was performed by a combination of gas
chromatography/mass spectrometry and nuclear magnetic
resonance, with in some cases comparison to commercial and
synthetic pure compounds. On the basis of vapor pressure
estimates of some typical ozonolysis products, it is
suggested that the initial atmospheric degradation of these
terpenes at ambient concentrations leads to gas phase
products. Also, a mechanistic suggestion is made concerning
a possible rearrangement of the Criegee intermediates
produced in the ozonolyses.

INTRODUCTION

The two principal pathways for the degradation of
hydrocarbons (HC's) in the atmosphere are reaction with ozone
(O_3) and reaction with hydroxyl radicals (OH's). A crude
estimate of the percentage of atmospheric degradation that
would result from reaction with O_3 and reaction with OH can
be made using background concentrations of these reactants,
and the second order rate constants that the terpenes display
with them. The results of such an estimate are given in
Table I for some typical terpenes.

161

TABLE I.   ESTIMATED PERCENT DEGRADATION
OF TERPENES IN THE ATMOSPHERE[*]

Terpene	% rxn O_3	% rxn OH
limonene	90	10
α-pinene	85	15
β-pinene	40	60
isoprene	30	70

[*]Assuming $[O_3] = 10^{12}$ mol cm^{-3} (40 ppb) and $[OH] = 4 \times 10^5$ mol cm^{-3}

For the more O_3-reactive terpenes, like limonene and α-pinene, the principal mode of degradation (assuming O_3 - OH constant concentration) would be by O_3. The slower O_3-reacting of the monoterpenes, β-pinene, and the hemiterpene isoprene, would be degraded principally by OH but with substantial degradation by O_3.

Arnts and Gay (1979), using Fourier transform-infrared (FT-IR) techniques were able to account for 4% to 17% of the reactive carbon as gas-phase carbon in nitrogen oxides (NO_x)-mediated photooxidation. The analysis was on species of 1 to 3 carbon atoms. In the particular cases of α-pinene and β-pinene, 9% and 6%, respectively, could be accounted for as gas-phase, small molecule carbon.

In work at Battelle, W. Schwartz (1974) looked at the photooxidation of α-pinene (10 ppm) in the presence of NO_x (3 ppm) in a large chamber. The aerosol produced was filtered, and accounted for 35% to 40% of the reacted α-pinene. Aerosol materials were analyzed by gas chromatography/mass spectrometry (GC/MS) techniques, using a chemical ionization mass spectrometer. The aerosols were first separated by extraction into acid, neutral, and basic fractions. The neutral and basic fractions were very similar. The acid fraction was analyzed by conversion of the sodium salt of the acids to methyl esters.

Most of the products were identified from the fragmentation pattern seen in the mass spectra obtained. One component, the cis-pinonic acid, was identified by direct comparison with a commercially available sample.

In all figures and tables in this paper, compounds that were identified only by mass spectral fragmentation patterns will appear in parentheses. Species that are not in parentheses were identified by comparison with commercially available materials or with synthetically produced materials.

The neutral aerosol fraction in Schwartz's work consisted of a number of compounds which are listed in Figure 1. Pinonaldehyde (note the question mark) has a molecular weight of 168. Schwartz did not specify which of the materials having molecular weight 168 actually was pinonaldehyde, but he suggested it was present because there were a number of species of that molecular weight with the appropriate kinds of fragmentation patterns.

In the GC/MS of the methyl ester fraction, Schwartz identified norpinonic acid, which is derived by cleavage of α-pinene at the pi bond with the loss of one carbon; cis-pinonic acid, which is derived by simple cleavage at the pi bond; and trans-pinonic acid. He suggested that the trans-pinonic acid was produced by photoisomerization of cis-pinonic acid. In order to separate the acid fraction, the collected aerosols were extracted with aqueous hydroxide.

In the present study, it was found that extraction of cis-pinonic acid with hydroxide equilibrates the cis-and trans-pinonic acids. There is an enolizable hydrogen in all of these species adjacent to the acetyl group on the ring. Under strongly basic conditions, this proton can be ionized and reattached with equilibration of the cis and trans isomers. Thus, the appearance of the trans-acid in Schwartz's samples may be an artifact.

The accepted mechanism of ozonolysis requires the formation of a five-membered ring species (a molozonide or a primary ozonide) that breaks down rapidly into a Criegee intermediate (often referred to as the Criegee carbonyl oxide, di-radical, zwitterion, etc., depending upon the point of view and on whether the work is done in gas or liquid) and a carbonyl compound. An outline of the process is given in Figure 2 (Bailey 1978).

In the liquid phase, various kinds of products can be obtained as indicated depending upon the conditions (chiefly solvent and substrate). Among the likely products are polymeric peroxides, in which the carbonyl oxide has polymerized or dimerized; secondary ozonides, which are a combination of the carbonyl oxide and some other carbonyl compound, in the mixture; and upon addition of alcohol across the carbonyl oxide in a 1,3 fashion, alkoxyhydroperoxides. These latter compounds are useful synthetically, as will be discussed below, because the alkoxyhydroperoxides can be reduced to carbonyl compounds and the ozonolysis reaction can be used for synthetic purposes.

Rearrangement products can be derived from any of the above, including the carbonyl oxide.

In the gas phase, as indicated, the source of radicals may be the Criegee intermediate, which appears to give rise to hydroperoxyl, hydroxyl, peroxy-alkyl, and alkoxyl radicals. There is a considerable amount of fragmentation in the production of these radicals.

O'Neal and Blumstein (1973) have also suggested that the primary ozonide, (shown in Figure 2) can also be the source of some of the observed products. They suggest the ozonide forms a di-radical which internally abstracts hydrogen and rearranges.

Ozonolysis of terpenes in the liquid phase has been studied for some time. In the early work, Harries and others (Simonsen 1953) used O_3 to locate the double bonds in the terpenes. The cis-pinonic acid was first observed in these α-pinene-O_3 reactions.

In the 1950's, some interest was displayed in cis-pinonic acid as a plasticizing agent. Attempts were made to develop commercial methods for its production, based on ozonolysis of α-pinene with an oxidative workup (Fisher and Stinson 1955).

Throughout this period, the terpenes have served as precursors of perfume agents; the formation of alcohols and ketones by oxidation-reduction sequences give pleasantly aromatic materials. Pinonaldehyde is relatively well known as a precursor to perfume-like materials (Kulesza and Kula 1975). The initial work in β-pinene-ozonolysis was again an attempt to locate the double bond (Simonsen 1953). The isolation of nopinone was a result of that work. No oxidative or reductive workup is required to isolate the nopinone, which is unusual in ozonolyses, and indicates that the secondary ozonide probably decomposes readily in aqueous solutions to give the nopinone. No product identification has been done with modern analytical methods on the gas phase α- and β-pinene-O_3 reactions. One report (Spencer et al. 1940) is available on the gas phase α-pinene-O_3 reaction in which the products were characterized largely by C-H analysis.

A. Acid Fraction

cis-pinonic acid

trans-pinonic acid

pinonic (norpinonic) acid

B. Neutral Fraction

cis-pinonaldehyde (?)

cis-norpinonaldehyde

C. Basic fraction — minor

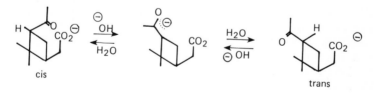

cis

trans

Figure 1.   Aerosol products in α-pinene/NO_x/hν (Schwartz 1974).

165

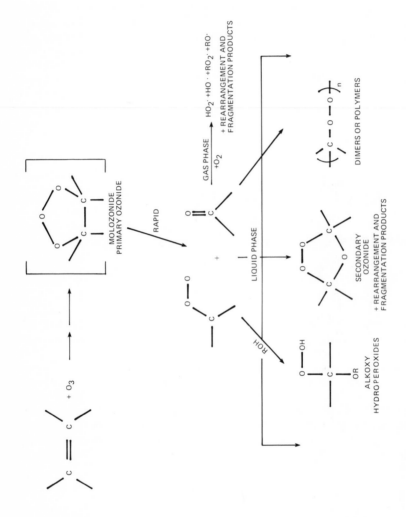

Figure 2. The mechanism for the ozonolysis of alkenes (Bailey 1978).

METHODS AND MATERIALS

Figure 3 is a schematic of the flow system used to generate product mixtures in the gas phase ozonolysis work that was performed in this study.

Ozone/air and terpene/air streams were mixed in the flow system. The effluent gases were then passed through a cold trap (at -78°C) which contained glass beads and methylene chloride. In one run, a 1 μm Teflon filter was attached after the trap to collect aerosol materials. After a certain amount of time, the reaction was stopped. The reaction line, cold trap, and in one case the filter, were extracted with methylene chloride, keeping each of the extracts separate. The data on weight and concentration for the α-pinene-O_3 runs are given in Table II.

The concentrations are high compared to ambient atmospheric conditions, but for reasons that will be mentioned later, they are probably still useful for purposes of product analysis. Total pressures were essentially one atmosphere; flow rates were on the order of 400-500 cc/min through the flow reactor.

The weights of materials collected are crucial. In the one run without the filter, only about a third of the terpene lost (α-pinene) could be accounted for as collectable product. In the case of the filtered run (Table II, Run B), the material balance was better, representing roughly 80% of the α-pinene lost (taking into account that the products can have one or more oxygens).

The flow reactor can also be used to look at the stoichiometry of the reaction by determining the α-pinene concentration in the line before the O_3 generator is turned on, and determining it again after the O_3 is mixed with the α-pinene. If the O_3 production of the generator has been calibrated reasonably well, the stoichiometric information can be found for the reaction.

In Table III, the results of the stoichiometry determinations for both α-pinene and β-pinene are given. The initial pinene concentration, initial O_3 concentration, initial ratio, and the stoichiometric consumption of pinene to O_3 are given. For both α-pinene and β-pinene the consumption ratios range from about 0.9 to 1.4.

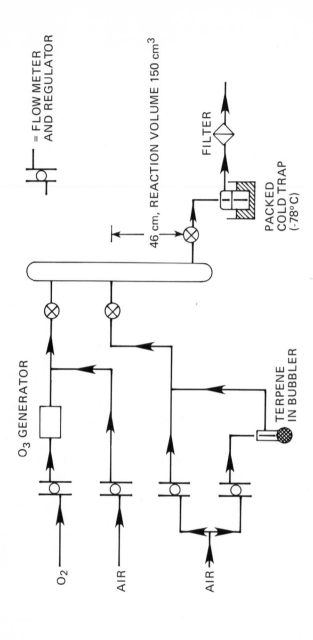

Figure 3. Schematic of the terpene–O_3 flow reactor system.

168

TABLE II. α-PINENE-O_3 REACTION DATA

	Run A	Run B
α-pinene	4.3 τ (4600 ppm)	(360 ppm)$_o$ (81 ppm)$_f$
O_3	3.0 τ (3900 ppm)	250 ppm
Total Press	753 τ (O_2)	O_2 (186 τ), N_2 (574 τ)
Flow	380 cc/min	563 cc/min
Run Time	220 mins	167 mins
Wt. Products		
Line	0.64 g	0.054 g (36%)
Trap	0.23 g	0.020 g (13%)
Filter	--	0.078 g (51%)*
TOTAL	0.87 g	0.152 g* (100%)
Wt. Terpene Consumed	2.1 g	0.145 g

*Extrapolated from the weight of filtered material
collected in 43 mins

The last listed α-pinene-O_3 run was done quite
differently. To a 4 ppm O_3/air mixture was added 16 ppm of
α-pinene in the long-path IR cell used in the FT-IR work done
by Arnts and Gay (1979). The FT-IR was used to determine the
concentration of the α-pinene after addition to the O_3. That
result, obtained at a factor of roughly 15 less in O_3
concentration than the lowest of the other determinations,
gives approximately the same stoichiometry as they do.

In the literature, Ripperton et al. (1972) presented
data from which one could extract an approximate α-pinene-O_3
consumption ratio. The ratio varied somewhat over the course
of their reaction, but was approximately 3 moles terpene:1
mole O_3. Grimsrud et al. (1975) did kinetic work on the
various terpenes, and determined a stoichiometry of about 1.1
to about 1.4 for the α-pinene-O_3 reaction which is in line
with the results presented here.

The stoichiometry suggests that for product purposes,
the reaction is the same at the high concentrations used here
and the low concentrations used by other workers. If there
were secondary radical reactions with the alkene at high
concentration, the stoichiometry would not be expected to be
approximately 1:1, and would not be the same as at lower
alkene concentrations.

169

TABLE III. STOICHIOMETRY OF THE TERPENE-O_3 REACTIONS

$[O_2]=190\tau$, $[N_2] = 570\tau$

A. α-Pinene

$(\alpha\text{-pinene})_0$ (ppm)	$(O_3)_0$ (ppm)	$\left(\dfrac{\alpha\text{-pinene}}{O_3}\right)_0$	$\dfrac{\Delta\alpha\text{-pinene}}{\Delta O_3}$
110	74	1.5	1.4
235	74	3.2	1.4
273	250	1.1	1.0
360	250	1.4	1.1
509	250	2.0	1.3
16.*	4.0*	4.0*	1.3*

B. β-Pinene

$(\beta\text{-pinene})_0$ (ppm)	$(O_3)_0$ (ppm)	$\left(\dfrac{\beta\text{-pinene}}{O_3}\right)_0$	$\dfrac{\Delta\beta\text{-pinene}}{\Delta O_3}$
89	62	1.4	1
127	62	2.0	1.3
305	306	1.0	0.91

C. Literature values

$(\alpha\text{-pinene})_0 + (O_3)_0$	$\dfrac{\Delta\alpha\text{-pinene}}{\Delta O_3}$	Reference
1 ppm + 0.2 ppm	~3	9
2-3 ppm + 0.2-0.8 ppm	1.1-1.4	10

*Long path IR determination

Sample Analysis

The liquid samples collected by extraction of the reaction line, trap, and filter were subjected to GC/MS and nuclear magnetic resonance (NMR) analyses. The two methods complement each other. The NMR does not subject the sample to high temperature and can therefore be used to check for artifacts resulting from decomposition in the injection port or column. The NMR will detect all the hydrogens in the mixture and not 'miss' the nonvolatile components. In addition, characteristic and distinct NMR absorptions can be used to confirm MS assignments and can also be used quantitatively.

Figure 4 shows the GC/MS total ion abundance spectrum of the filtered material from the α-pinene–O$_3$ reaction on an SE-30 capillary column programed from 90°C to 250°C. The peaks assigned in Figure 4 were identified either by comparison with authentic samples (no brackets) or by MS assignment (in brackets).

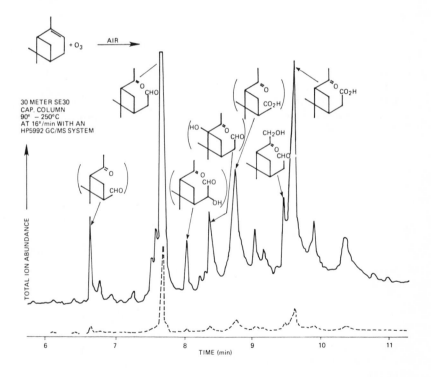

Figure 4. GC/MS total ion abundance spectrum on a β-pinene/O$_3$/air product mixture.

171

The acids were difficult to analyze. At low concentrations in the methylene chloride solutions, materials were lost on the column after injection. Injection of more concentrated samples resulted in good spectra.

The cis-pinonic acid was a commercial sample; pinonaldehyde and the OH substituted pinonaldehyde (I) were synthesized as described below. In addition, there were a number of species that were tentatively identified on the basis of mass spectra. The cis-pinonic acid peak and the peak labeled as nopinonic acid were both extractable by bicarbonate.

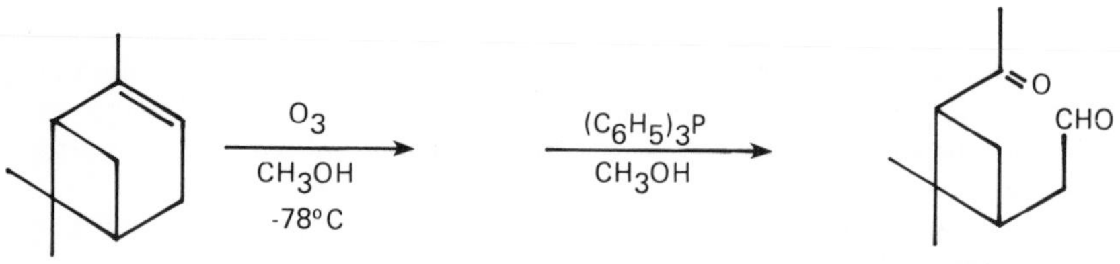

$$\xrightarrow[\substack{CH_3OH \\ -78^\circ C}]{O_3} \xrightarrow[CH_3OH]{(C_6H_5)_3P}$$

pinonaldehyde

Pinonaldehyde and (I) were synthesized in the two step process outlined below. Ozonolysis in alcohol solvents results in the production of alkoxyhydroperoxides, which are readily reduced to carbonyl compounds by triphenylphosphine without isolation of any peroxidic materials. The process is called reductive ozonolysis and is a widely applied reaction (Bailey 1978).

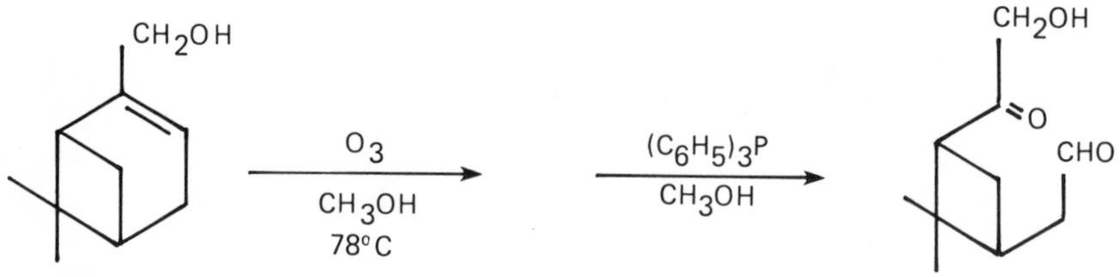

$$CH_2OH \quad \xrightarrow[\substack{CH_3OH \\ 78^\circ C}]{O_3} \xrightarrow[CH_3OH]{(C_6H_5)_3P} \quad CH_2OH$$

myrtenol

(I)

The NMR and MS results are entirely consistent with the structures assigned. The mass spectra and the retention times of the synthetic materials are identical to those found for the indicated components of the gas phase product mixture.

In Figure 5, the identified products and the quantitative data on the known components of the α-pinene-O$_3$ product mixtures are summarized. All three fractions (reaction line, trap, and filter) showed virtually the same product distributions. All the filtered material was extractable into methylene chloride, none remained on the filter. The material collected could be fractionated into an acid fraction and a neutral fraction. About 45% of the mixture was acidic; about 55% was neutral. In the acid component, the pinonic acid could be identified in both the NMR and the GC/MS. Based on the NMR integration of methyl signals, about 7% nopinonic acid was estimated to be present. Some additional acidic material was also present (not the trans-pinonic acid).

In the neutral fraction, the pinonaldehyde accounts for about 15% of the total material and (I) for about 7%. The various other materials were shown by GC/MS to be each 5% to 10% of the total.

Note that the pinonaldehyde was also observed in the FT-IR experiment described earlier for the stoichiometry determination. If the residual α-pinene is subtracted from the spectrum obtained for the product mixture, what is left is a spectrum that shows the presence of the pinonaldehyde. The pinonaldehyde, therefore, is seen not only in the GC/MS and the NMR, it is also seen directly in the gas phase in the FT-IR experiment.

Table IV summarizes the data for the reactions of O$_3$ with β-pinene. Again, the second reaction was performed with a filter on the effluent gases.

As in the α-pinene case, it was found that each of the three different fractions (the material isolated from the line, the trap, and the filter) showed the same composition.

As before, the analyses were performed using a combination of GC/MS and NMR techniques. Figure 6 shows a GC/MS total ion abundance spectrum under the same analytical conditions as used for the α-pinene-O$_3$ product analysis. Myrtenol (pin-2-en-10-ol) was available commercially (Aldrich Chemical, Inc., Milwaulkee, Wisconsin); nopinone, 3-hydroxynopinone, and 3-ketonopinone were synthetic materials. The nopinone was synthesized by ozonolysis of β-pinene in a manner similar to that already described. The 3-hydroxy-and 3-ketono-pinone were synthesized as indicated in Figure 7 (Jefford et al. 1973, Corey and Suggs 1975). The assignment of the broad peak at 7.8 mins in Figure 6 to the β-pinene secondary ozonide was based on comparison of the gas phase results to the liquid phase results. The broad peak is actually due to nopinone, which results from decomposition of the secondary ozonide on the column. In the NMR, it is possible to see the characteristic secondary ozonide peak at 5.2 disappear with the addition of triphenylphosphine.

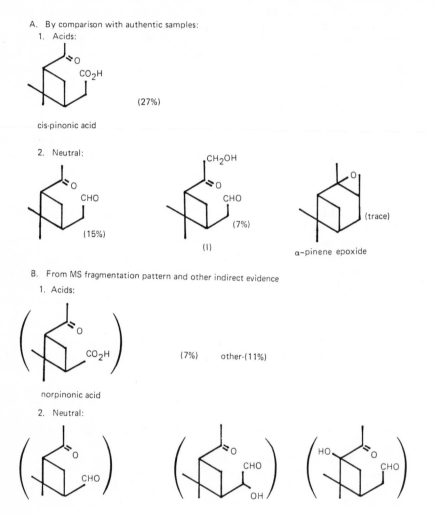

A. By comparison with authentic samples:
  1. Acids:

  cis-pinonic acid (27%)

  2. Neutral:

  (15%)   (1)   α-pinene epoxide (trace)   (7%)

B. From MS fragmentation pattern and other indirect evidence
  1. Acids:

  norpinonic acid   (7%)   other-(11%)

  2. Neutral:

Figure 5. Products observed in the gas phase α-pinene–O$_3$ reaction.

TABLE IV. β-PINENE-O₃ REACTION DATA

	Run A	Run B	
$[O_3]$	270 ppm	179 ppm	
[β-pinene]	440 ppm	(275 ppm)o	(90 ppm)f
$[O_2], [N_2]$	170,590	185,575	
Flow Rate (cm³/min)	380	590	
Time (min)	215	266	
Wt. Products			
Line	.012 g	.0857 g	
Trap	.011 g	.0175 g	
Filter	--	.0407 g*	
TOTAL	0.023 g	0.144 g*	
Wt. of Terpene Consumed	0.155 g	0.161 g	

*Extrapolated from the weight of filtered material collected in 70 mins

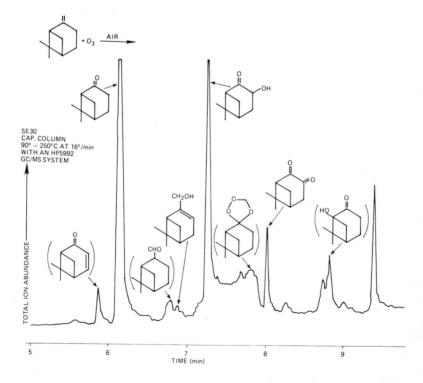

Figure 6. GC/MS total ion abundance spectrum on a β-pinene/O₃/air product mixture.

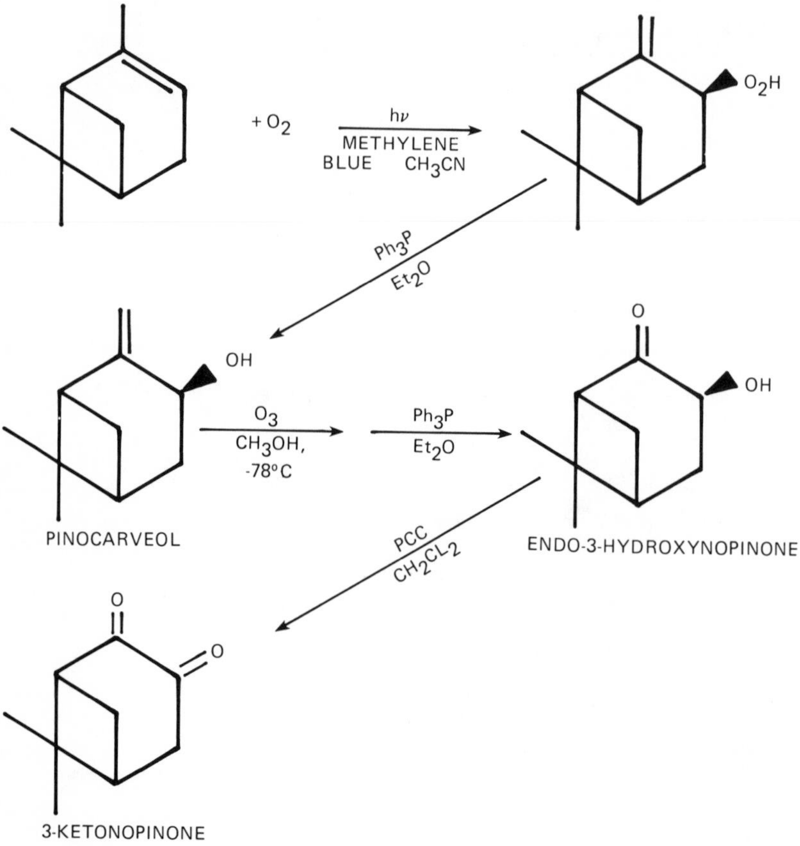

Figure 7.   Synthesis of 3-hydroxynopinone.

The material identified as 3-hydroxynopinone was
initally assigned on the basis of its mass spectrum and the
NMR spectra. The NMR of the product mixture showed a small
absorption, a doublet or doublets, in the 4.2-delta region,
which is assignable to a hydrogen next to a hydroxyl group.
The splitting pattern, the doublet of doublets, was
assignable to a hydrogen in a cyclic system coupled to an
equatorial and axial hydrogen on an adjacent carbon.
Subsequent synthesis (Figure 7) confirmed the assignment.
The synthetic material showed the identical mass spectrum and
retention time as that produced in the gas phase ozonolysis
reaction. This may indicate that the gas phase product is
the particular stereoisomer shown in Figure 7 (endo-3-hydro-
xynopinone) since the synthesis is stereospecific (Jefford et
al. 1973).

Figure 8 summarizes the products identified in the
β-pinene-O_3 reaction and the approximate percentages of the
major products. The products identified by comparison with
authentic samples account for roughly 50% of the product
mixture, and a large number of other products account for the
other 50%. These latter compounds, however, differ very
little in molecular weight (and presumably volatility) from
the identified materials. The GC retention times of these
products are similar, and their mass spectra show that they
are mono- and di-oxygen derivatives of the terpenes. The
GC/MS and NMR results indicate that the products are not high
molecular weight polymeric materials.

With information about the product composition,
reasonable estimates of the physical properties of these
mixtures can be made. Table V lists estimates of the vapor
pressures at 25°C for some of the products of the terpene
ozonolyses. For two of the compounds, there are boiling
point data on the material itself. For the other compounds,
model compounds were chosen that mimic fairly well the
compound in question as to molecular weight and functional
groups.

The vapor pressure estimates were based on the linear
relationship between the log of vapor pressure and 1/T, with
empirical estimates (based on the type of compound) of the
enthalpy and entropy of vaporization. The method is that of
Hass and Newton (1972) and is a standard method for boiling
point extrapolation estimates. The accuracy of the estimates
is probably good enough for our purposes.

A. By comparison with authentic samples:

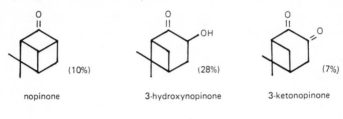

nopinone (10%)   3-hydroxynopinone (28%)   3-ketonopinone (7%)

myrtenol   pinocarveol

B. From MS fragmentation or other indirect evidence:

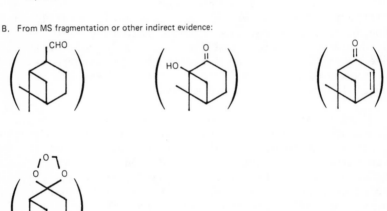

Figure 8.   Products of the gas phase β-pinene-O$_3$ reaction.

TABLE V.   VAPOR PRESSURE ESTIMATES

Compound	Model	B.P.(°C)	V.P. Estimated At 25°C (ppm)
nonpinone	–	77 (12 mm)[1]	700
3-hydroxynopinone	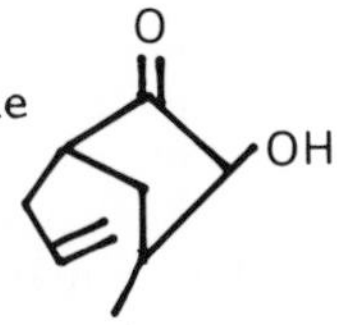	85 (0.6 mm)[2]	1.0
cis-pinonaldehyde	–	85 (1. mm)[3]	20
cis-pinonic acid	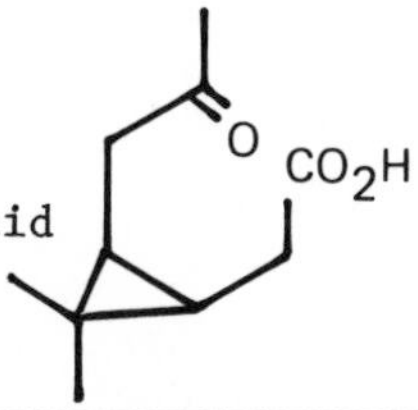	146 (2. mm)[4]	0.1

[1](Jefford et al. 1973)
[2](Beilsteins 1948)
[3]This work
[4](Noves et al. 1961)

DISCUSSION

Cis-pinonic acid is a solid at 25°C. An accurate estimate of its vapor pressure cannot be based on the model for the boiling point of a liquid, because the phase change lowers the vapor pressure. All that can be done is to put an upper limit on the cis-pinonic acid vapor pressure based on a liquid phase model.

All of the numbers are high compared to those that might be found under atmospheric conditions. The concentrations of the terpenes suggested earlier limit the concentration of the ozonolysis products. The products cannot be in any greater

179

concentration than the original terpenes; a factor of 1,000, or more, less than these vapor pressure estimates. It is therefore likely that the products of NO_x/air mediated photooxidation (largely ozonolysis) would remain in the gas phase. While some removal may occur by selective absorption on particulate matter, it is unlikely given the vapor pressures that this could account for significant amounts of material.

Smog chamber modeling of these systems can be misleading. The terpene concentrations are usually in the 0.1 ppm to 1.0 ppm range, at which pressure the products may come out of the vapor phase and therefore not mimic atmospheric conditions. If the products are removed from the gas phase, it can be assumed that they will not be available for further photodegradation. Chamber studies may well result in more aerosol formation than would be observed under natural conditions.

Ozonolysis results essentially demonstrate only the first step in any air degradation of the terpenes. Probably, more highly oxygenated materials would be found under natural conditions, particularly if the ozonolysis products are in the vapor phase. The ozonolysis products should be further degraded in the atmosphere by photolysis of the carbonyls and by hydroxyl radical abstractions of hydrogens. Because these materials have more unfunctionalized C-H bonds than are encountered in propene or butene ozonolysis products, they should be less reactive than those products on a per carbon basis. The reactivity of the terpenes in the atmosphere can be modeled mathematically by using a mixture of alkanes and alkenes to create the correct mix of functionalized carbons.

The ozonolysis products in this study were easily detected because they were in high concentrations relative to what would be available in atmospheric sampling. Detecting these materials under atmospheric conditions requires a great deal of care because the products are highly polar and difficult to analyze. It may, however, be possible, using appropriate techniques, to find these compounds in the atmosphere in rural areas.

The product mixtures seem to contain a number of alpha hydroxylcarbonyl compounds. In the β-pinene ozonolysis, more than twice as much of the hydroxyl substituted material is present than the conventional ozonolysis product, nopinone. In the α-pinene case, one material positively identified has an alpha hydroxycarbonyl compound (I), and several of the other components tentatively identified also have that kind of structural feature. This suggests that there may be a mechanistically important pathway which leads to this type of product. Hydroxyl radical abstraction alpha to the carbonyl group is certainly a possibility.

Alpha hydroxycarbonyl compounds were also observed by Story and Burgess (1967, 1968), in their work on the ozonolysis of tetramethylethylene in the liquid phase. They found hydroxyacetone and a product resulting from its presence. Figure 9 presents their rationalization of that result; it indicates that approximately 11% of their products could be accounted for on the basis of the hydroxyacetone. The hydroxyacetone is, of course, the alpha hydroxycarbonyl analogue of the corresponding terpene ozonolysis products. The suggestion is made that the Criegee intermediate rearranges to give the hydroxyacetone because the carbonyl oxide of acetone is otherwise unreactive in ozonolysis reactions; that is, it doesn't tend to form secondary ozonides and it is not trapped by solvent (pentane). Thus, because it is long-lived it has a path open to it that is not available to other less persistent carbonyl oxides.

Under gas phase conditions at low concentrations, the carbonyl oxides produced on ozonolysis of either α-pinene or β-pinene are, like acetone carbonyl oxide, highly substituted and less subject to rapid trapping than other carbonyl oxides produced from similar alkenes. The terpene carbonyl oxides might be expected to imitate acetone carbonyl oxide and undergo rearrangement to give alpha hydroxycarbonyl compounds. This is illustrated below for β-pinene. This suggestion, however, is quite speculative and certainly will require further work for substantiation.

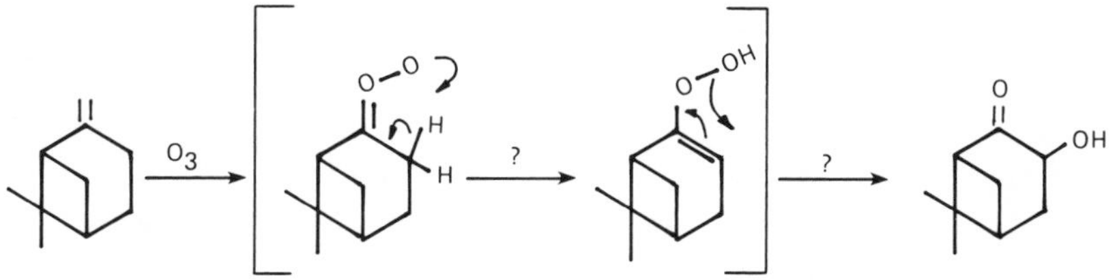

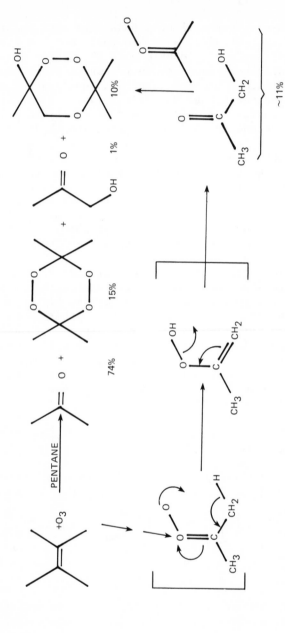

Figure 9. The ozonolysis of tetramethylethylene (Story and Burgess 1967, 1968).

REFERENCES

Arnts, R. and B. Gay. 1979. Photochemistry of Some
Naturally Emitted Hydrocarbons. EPA-600/3-79-981. U.S.
Environmental Protection Agency, Research Triangle Park,
North Carolina. 128 pp.

Bailey, P. S. 1978. Ozonation in Organic Chemistry. Vol.
I. Academic Press. Inc., New York. 272 pp.

Beilsteins Handbuch der Organischen Chemie. 1948. 4th ed.
2nd supplement, Springer-Verlag, Berlin, 8:16.

Corey, E. and J. Suggs. 1975. Pyridinium chlorochromate-
efficient reagent for the oxidation of primary and secondary
alcohols to carbonyl compounds. Tet. Letters. 1975:2647.

Fisher, G. S. and J. S. Stinson. 1955. Pinonic acid
preparation by the ozonolysis of alpha-pinene. Ind. and Eng.
Chem. 47:1569-1572.

Grimsrud, E., H. Westberg, and R. Rasmussen. 1975.
Atmospheric reactivity of monoterpene hydrocarbons, NO_x
photooxidation and ozonolysis. Inter. J. Chem. Kin., Sym 1,
pp. 183-195.

Hass, H. and R. Newton. 1972. Correction of boiling points
to standard pressure. In: Handbook of Chemistry and
Physics, 53rd edition. CRC Press, Cleveland, Ohio. pp.
D144-145.

Jefford, C., A. Boschang, R. Moriarty, C. Rimbault, and M.
Laffer. 1973. The reaction of singlet oxygen with alpha and
beta-pinenes. Helv. Chem. Acta. 56(6):2649-2659.

Kulesza, J. and J. Kula. 1975. Ozonolysis of terpenes for
perfurm synthesis. Riechst., Aromen, Koerperpflegen.
25(11):317-321.

Noves, Y. and A. V. Grampoloff. 1961. Etudes sur les
metieres vegetals volatiles CLXXV sur les produit obtenus a
partir de l'ozonolyse (+)-Δ_3-carene. Helv. Chem. Acta.
44:637.

O'Neal, H. E. and C. Blumstein. 1973. A new mechanism for
gas phase ozone-olefin reaction. Int. J. Chem. Kin.
5:397-413.

Ripperton, L., H. Jeffries, and O. White. 1972. Formation of aerosols by reaction of ozone with selected hydrocarbons. In: Photochemical Smog and Ozone Reactions, R. F. Gould, ed., ACS Advances in Chemistry Series, 113:219-231.

Schwartz, W. 1974. Chemical Characterization of Model Aerosols. EPA-650/3-76-085. U.S. Environmental Protection Agency, Research Triangle Park, North Carolina, 80 pp.

Simonsen, J. L. 1953. The Terpenes. Vols. I and II. Cambridge University Press, London.

Spencer, C., W. Weaver, E. Oberright, H. Sykes, A. Barney, and A. Elder. 1940. Ozonization of organic compounds. J. Org. Chem. 5:610-617.

Story, P. and J. Burgess. 1967, 1968. Ozonolysis evidence for carbonyl oxide tautomerization and for 1,3-dipolar addition to olefins. J. Amer. Chem. Soc. 89(22):5726-5727 (1967) and 90(4):1094 (1968).

DISCUSSION OF PRESENTATION

GRAEDEL: The liquid phase work that you refer to, is that normally done in organic solvents?

HULL: Yes. The synthetic work is done in methanol as solvent. If one doesn't want to trap the materials as alkoxyhydroperoxides, the ozonolysis can be performed in hydrocarbon or chlorinated hydrocarbon solvents. Story's work was in pentane, but CCl_4 is often used.

WALKER: Was there any evidence of polymeric products?

HULL: There was none in this work. The comparison of the GC/MS results (in which polymeric material might not be detected) with the NMR results (which "sees" all the hydrogens) shows that all the products are accounted for as volatile materials.

OLLISON: Do some of these things have nice odors, where you're postulating very low vapor pressures? Have you sniffed them, do they—

HULL: Yes, they do. You can smell all of these. The cis-pinonic acid doesn't have an odor. I can't really detect anything from that. But the others all do. Some of them are very nice.

BUFALINI: There's one thing that still bothers me. In keeping with the Gay-Arnts findings, if indeed the vapor pressure is that high, why would the material, even if it is polar, go to the walls and not be re-evaporated?

HULL: Well, I was looking at Murray's results and I would say they are consistent with re-evaporation. The constant rate of production of carbon monoxide after an an initial burst seems to me consistent with the consumption of some material in the gas phase and a constant flux of material from the walls, that is being then photooxidized to give you the carbon monoxide.

BUFALINI: Yes, but your vapor pressures are so much higher. You're talking about parts per million--for the vapor pressures. The material partly comes off the walls; but the rate of formation isn't that great.

HULL: These are, of course, vapor pressure estimates for pure materials. Obviously you don't have pure materials in the product mixtures. One would expect some vapor pressure lowering based on the mole fraction present although this is probably limited to a factor of 10 less. At the higher vapor pressures encountered in the chambers, one would expect more selective absorption on particulate matter than under atmospheric low vapor pressure conditions.

BUFALINI: Unless they can be observed on the surface.

HULL: Unless there can be some kind of imitation of a bonding to the surface.

JEFFRIES: To summarize the carbon split in your system, my impression is that you had a third of the material in the gas phase, a third of the material on the walls, and a third of the materials in the--

HULL: I would summarize it differently. I would say I had virtually all products as aerosols. The analysis of the material indicated that all the different fractions were approximately the same in composition. It was simply that some of the aerosol deposited in the line, some of it would deposit in that cold trap, and some would be caught by the filter. The cold trap, incidentally, accounted for the least amount of material. That is, it was the least efficient way to collect material in my system. I suspect that's due to the fact that most of this material is aerosol and when you try and trap it out in a cold trap, it doesn't trap out, it simply goes right through.

The line material, in the α-pinene case, plus the filtered material accounted for 70% to 80% of the collected material. Those have to be aerosol.

I would say, except for maybe 10% that corresponds to the small gas phase molecules, such as those seen by Arnts and Gay, all the rest is aerosol product under the high reactant concentrations used in this study. They may not be aerosol at lower reactant concentrations but they're these compounds.

KAMENS: But at lower concentrations it is possible to get a lot of these species in the gas phase, because of vapor pressure, isn't it?

HULL: I think that's correct. That may have something to do with Stevens' results from collecting the Smoky Mountains materials.

W.E. WILSON: The problem of why you find materials in the aerosol phase, when the vapor pressure of pure material is much too high for that to exist, has puzzled us for a long time. But I just want to remind people that we have done high resolution GC measurements, and other people have done GC/MS as well, and we see many of these compounds, including things like the di-acids --very short chain di-acids--in ambient aerosols, even though their vapor pressures are much, much too high for them to be in ambient aerosols. So there is some mechanism that reduces the vapor pressure. Whether it's a lot of polar compounds getting together, or whether it's some species forming a substrate--sulfuric acid, carbon, aerosol, or whatever--they do appear in the aerosol phase in the ambient air. So you should find it in a smog chamber.

BUFALINI: Is this published in the haze reports? Or is this something new?

W.E. WILSON: This is published, at least in EPA reports.

18. THE IMPACT OF ALPHA-PINENE ON URBAN SMOG FORMATION:
AN OUTDOOR SMOG CHAMBER STUDY

Richard M. Kamens. Department of Environmental Sciences
and Engineering, University of North Carolina, Chapel
Hill, North Carolina 27514

ABSTRACT

To determine the effects on ozone production, carbon
supplied by α-pinene was either added to or used to replace
20% of the carbon contained within an urban-like mixture of
eleven hydrocarbon species in a large outdoor smog chamber.
Although α-pinene is somewhat more reactive than the
hydrocarbon mixture, direct replacement of ~20% carbon
results in no change in overall reactivity, as indicated by
NO_x and O_3 behavior. The addition of 20% extra carbon as
α-pinene produces similar changes to increasing carbon
concentration of the hydrocarbon mixture by 20%.

If α-pinene and ozone are first reacted, their products
do not impart any unusual reactivity to a hydrocarbon-NO_x
smog system. In one experiment a loss of reactivity was
observed, as measured by ozone production. Aerosol
measurements made when α-pinene and ozone reacted indicate
that the majority of particles that form is in the 0.1 μm
diameter range; however, particles >0.6 μm diameter contained
most of the mass. Carbon accountability in the product-
condensed-phase was estimated to be 25-90%, but a very high
uncertainty is associated with this estimate.

INTRODUCTION

In recent years the relative importance of natural
hydrocarbons (HC's) in both urban and rural smog formation

has been a controversial topic. Attempts to relate natural HC emission rates with natural HC ambient levels have met with varying degrees of success. For example, Westberg and Holdren (1976), using Zimmerman's (1977) experimentally-determined, plant-specific emission rates, were able to predict an average terpene concentration of 16 ppbC within a 20 m high pine forest. This figure was compatible with the ambient values they also measured. When Zimmerman's plant-specific emission rates were later used in conjunction with biomass and other factors for the Tampa Bay-St. Petersburg area (1978), the final computation suggested that 68% of all HC emissions were of natural origin. But these estimates appeared to be at variance with EPA measurements (Lonneman et al. 1978) that showed almost all HC's in the Tampa Bay-St. Petersburg plume to come from anthropogenic sources.

Although much discussion and debate has resulted from the discrepant Tampa Bay emissions and ambient measurements, one is forced to consider that experimentally-determined plant emissions and biomass estimates may produce regional emission estimates that are somewhat high. At the same time, anthropogenic emission estimates for the same area may be optimistically low. Given the many assumptions used to calculate both natural and anthropogenic emissions, a considerable error in estimates (Lonneman and Bufalini 1979) is certainly possible.

To further complicate matters, a statistical analysis (Sandberg 1978) of total HC (THC) and O_3 data from the San Francisco Bay area attempted to relate increased natural HC emissions with increased summer O_3 concentrations. The increase in THC was alleged to result from abnormally wet winters that would cause greater plant growth, higher emissions, and subsequently more O_3. Although this THC data had no accompanying methane or detailed natural HC information, extremely wet winters could cause an increase in biogenic decay and, hence, produce greater local methane emissions (which would be directly reflected in THC readings). But Miller et al. (1979) analyzed similar San Francisco Bay data sets and could not verify the correlations observed in the original Sandberg study.

The present study was conducted, using a large outdoor smog chamber, to address relevant questions about the photochemical activity of natural HC's. Since certain urban emission inventories used by modelers contain large fractions of natural HC's, these experiments were to focus on the possible contribution of natural HC's to urban O_3 formation and the photochemical reactivity of the products formed from α-pinene and O_3.

EXPERIMENTAL

The experimental protocol employed for conducting runs in the UNC dual outdoor chamber has been described in detail by Jeffries et al. (1975). In this study α-pinene was used to represent natural terpenoid HC's, and was injected at predawn hours into one compartment of the outdoor chamber. An eleven component urban-like HC mix (70% paraffins and 30% olefins, by carbon (see Table I) was then added to both chamber halves such that the effect of α-pinene on O_3 formation could be observed.

In most of the runs α-pinene did not make up more than 20% of the total nonmethane HC (NMHC) mix; α-pinene was either substituted for carbon in the mix so that both sides of the chamber had the same initial carbon concentration, or so that one side had about 20% more carbon than the other. Comparisons with the HC mixture on one side and α-pinene only on the other were also performed.

In all experiments, the injected morning levels of NO and NO_2 were the same. On a few occasions, early morning α-pinene was allowed to react in the dark for approximately 2 h with low levels of O_3, then the HC mix and NO_x were added. The experiments were then carried out in the sun in a fashion similar to that described above. Summary data for the experiments of interest are given in Table II and sample smog profiles for both an α-pinene carbon substitution and addition to the mix are given in Figures 1 and 2.

RESULTS AND DISCUSSION

It is useful in discussing the reactivity of α-pinene within an urban-like mix environment to compare the O_3 yields from similar carbon concentrations of mix and α-pinene that both contain the same initial NO_x. Such an experiment was performed on August 18, 1979, in which approximately 1.3 ppmC of mix was added to one chamber half and 1.3 ppmC α-pinene to the other. Both sides started with 0.19 and 0.05 ppm of NO and NO_2. The side with α-pinene produced a little more O_3 (0.39 ppm) than the side with mix (0.32 ppm O_3) and did have a somewhat faster rate of oxidation of NO to NO_2.

TABLE I. 1979 UNC SIMULATED URBAN HC MIX

| Class/Compound | Carbon Fraction | | Mole Fraction |
	UNC Analysis[*]	Matheson Analysis	UNC Analysis
Paraffins			
isopentane	0.1484	.1270	0.1273
n-pentane	0.2531	.2535	0.2411
2-methyl-pentane	0.0996	.1027	0.0712
2,4-dimethyl-pentane	0.0864	.0791	0.0528
2,2,4-dimethyl-pentane	0.1202	.1111	0.0644
Subtotal	0.7077	.6734	0.5568
UNC Average Carbon Number = 5.89			
Olefins			
1-butene	0.0254	0.0301	0.0272
cis-2-butene	0.0313	0.0297	0.0335
2-methyl-butene	0.0347	0.0393	0.0298
2-methyl-2-butene	0.0317	0.0393	0.0272
ethylene	0.1167	0.1343	0.2504
propylene	0.0524	0.0539	0.0750
Subtotal	0.2922	0.3266	0.4431
UNC Average Carbon Number = 2.78			
TOTAL	0.9999	1.000	0.9999

[*]UNC analysis is based on an average of many determinations of what actually appears in the chamber

TABLE II. α-PINENE AND URBAN MIX EXPERIMENTS
UNC DUAL SMOG CHAMBER RUNS[*]

Date	Initial HC (ppm)	NO_x (ppm)[†]	O_3 Max (ppm)
Substitution of α-pinene for mix			
8/18/79	1.31 α-pinene	0.241	0.391
	1.33 mix	0.240	0.323
8/31/79[‡]	1.26 mix	0.351	0.063
	1.24 α-pinene	0.354	0.160
9/01/79	0.67 mix + 0.55 α-pinene	0.219	0.287
	1.22 mix	0.208	0.258
6/10/79	1.60 mix + 0.34 α-pinene	0.302	0.252
	2.03 mix	0.299	0.248
7/06/79	2.09 mix + (0.2 α-pinene crossover)	0.350	0.192
	2.08 mix + (0.2 mix crossover)	0.345	0.223
Addition of α-pinene to mix			
6/09/79	2.12 mix 0.43 α-pinene	0.301	0.467
	2.14 mix	0.300	0.388
7/05/79	2.13 mix + 0.30 α-pinene	0.359	0.468
	2.15 mix	0.354	0.371
Increased carbon effect of mix			
7/14/79	2.24 mix	0.315	0.340
	2.65 mix	0.309	0.462
7/04/79[‡]	2.22 mix	0.348	0.201
	2.73 mix	0.342	0.257
Early morning reaction of O_3 and α-pinene			
9/11/79	1.62 mix + (0.07 O_3 + 0.42 α-pinene)	0.392	0.094
	2.02 mix	0.392	0.145
10/02/79	1.93 mix	0.201	0.264
	1.51 mix + (0.06 O_3 + 0.52 α-pinene)	0.205	0.265
7/24/79	1.62 mix (0.08 O_3 + 0.41 α-pinene)	0.414	0.663
	2.03 mix	0.420	0.670
7/13/79	2.25 mix	0.357	0.41
	2.25 mix (0.18 O_3 + 0.38 α-pinene)	0.355	0.57
Matched mix and NO_x			
7/03/79	1.82 mix	0.337	0.411
	1.82 mix	0.333	0.422
7/28/79	2.36 mix	0.436	0.241
	2.28 mix	0.428	0.226
6/19/79	4.73 mix	0.307	0.304
	4.71 mix	0.301	0.320
6/05/79	4.73 mix	0.403	0.863
	4.71 mix	0.400	0.865
10/04/79	2.72 mix	0.165	0.294
	2.74 mix	0.165	0.287

[*]Each date shows initial conditions and O_3 max for both chambers
[†]20% NO_2
[‡]Overcast

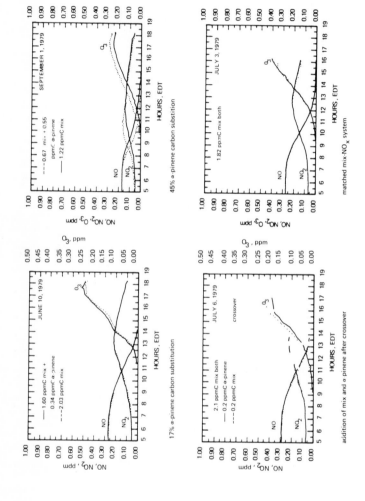

Figure 1. Effect of α-pinene carbon replacement in an urban-like hydrocarbon mix-NO$_x$ system.

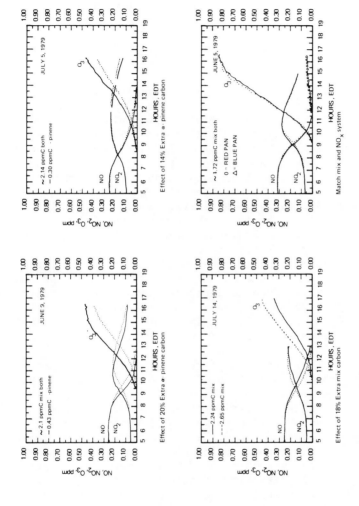

Figure 2. Effect of α-pinene carbon replacement in an urban-like hydrocarbon mix–NO_x system.

Alternatively, Gay and Arnts (1977) compared the O_3 yields from various terpenes and O_3 in an indoor chamber with constant light. The α-pinene HC/NO_x ratio (NO_x = 0.33 ppm) was optimized for maximum O_3 production (30:1 ppmC/ppm), yet even under these conditions α-pinene produced one third to one half the O_3 that propylene did. From the standpoint of NO oxidation and O_3 production, the photochemical reactivity of α-pinene by itself, then, is slightly greater than that of the UNC mix and much lower than that of propylene.

Substitution of α-Pinene Carbon for Mix Carbon

Alpha-pinene carbon was used to replace the mix carbon to see if any changes in NO_x or O_3 behavior could be observed. Direct replacement with 45% α-pinene (by carbon) was performed on September 1, 1979 (Table II). The side of the smog chamber containing α-pinene plus mix produced a little more O_3 (0.28 ppm compared to 0.25 ppm) and reached NO-NO_2 crossover about 1 h sooner than the 100% mix side. When replacement was lowered to 17% (June 10, Figure 1) the impact of the slightly more reactive α-pinene was almost completely lost; both sides had nearly identical NO, NO_2, and O_3 time profiles. It has been suggested that if α-pinene were added after crossover to a dual run which had the same mix and NO_x in both sides, then an α-pinene effect might be observed. This was done on July 6; 0.2 ppmC α-pinene and 0.2 ppmC mix were added to separate sides of a matched run of 2.1 ppmC and 0.35 ppm NO_x when crossover occurred. As shown in Table II and Figure 1, a slight but negligible reduction in O_3 production was observed in the α-pinene side.

Kamens et al. (1979) and Jeffries (1980) have proposed that 20% carbon substitution of most HC's into a complex urban-like HC mixture does not greatly alter the resultant NO_x or O_3 behavior. Jeffries et al. (1975) presented data for aromatic substitution as well as for more involved replacement of various HC within a mixture. All of these experiments suggest that noticeable changes in the overall reactivity of the mix occur only when compounds of extremely low or high reactivity are substituted into the mix. The implications of these findings may be of value to control agencies which consider HC substitution as a way to lower the reactivity of urban emissions and thereby a means for oxidant control.

Addition of Extra α-Pinene Carbon to a HC Mixture

When 20% extra α-pinene carbon was added to a matched mix dual chamber run, the side with α-pinene generated about 20% more O_3. Similar experiments (Table II) were then run with the HC mix, except that one chamber side had ~20% more mix than the other. As can be seen (Figure 2), both sides of these systems exhibit similar behavior and, hence, the change associated with the addition of 20% extra α-pinene carbon seems to be mostly that of an increased carbon effect.

Reaction of O_3 and α-Pinene Prior to the Injection of Mix

In some experiments low concentrations of O_3 (0.068 ppm) and α-pinene (0.4 ppmC) were added to the chamber and allowed to react for 1-2 h in the dark before the addition of either mix or NO_x. These tests were designed to determine if the product of this reaction in either the vapor or aerosol phase would influence the net O_3 production from a carbon mix/NO_x system. An experiment of this type was conducted on September 11, 1979. Formation of condensation nuclei was observed about 15 min after the addition of O_3 to α-pinene in the dark.

The mix and NO_x were finally added to both sides of this experiment so that each side had the same total carbon (i.e., initial α-pinene + mix = mix) and the same initial NO and NO_2 (after O_3 had been titrated by the NO injection). Hence, this run was similar to that on July 10, except the O_3 and α-pinene had an opportunity to react before the addition of either mix or NO_x had taken place. (Recall that in the June 10 experiment [Figure 1] both sides generated the same O_3). In this experiment the side with reacted α-pinene generated less O_3 (Figure 3).

The results of two other runs in which α-pinene and O_3 were first added to the chamber are uncertain because of missing data or poor meteorological conditions. They do, however, tend to support the results of the September 11 run. On the other hand, runs on July 13, 1979, and October 2, 1979, (Figure 3) did <u>not</u> show the reduced O_3 effect produced by first reacting O_3 and α-pinene. These experiments behave in the same way as the substitution and addition runs in Figures 1 and 2. Hence at this point one can say that all of the runs in which O_3 and α-pinene were first reacted do not show any more reactivity than those in which α-pinene was not first reacted, and in certain cases do show less O_3 production.

One might speculate that if reacting O_3 and α-pinene first leads to less O_3 production, then the products of α-pinene and O_3 may stabilize in the dark (i.e., would not be available for reaction) and lose some of their photochemical activity. On the other hand, when α-pinene is not first reacted with O_3, it participates as does any HC in a morning smog system. In other words, the products which result from either OH or O_3 attack will subsequently lead to NO oxidation before these products have an opportunity to stabilize.

One might further conjecture that the nature and amount of aerosol vs. vapor phase product may also influence the O_3 yield from the subsequent smog system. If different amounts of aerosol and products were generated in the above experiments, then perhaps the relative importance of other reactions like aerosol radical scavenging could be shifted, partially accounting for the apparent inconsistencies of the experiments in Figure 3. Unfortunately, chemical characterization of the products from ppm levels of α-pinene and O_3 is very difficult, with reasonable success at high concentrations only very recently achieved (Hull 1980). In two selected experiments, physical measurements of the aerosol were made in the UNC chambers, which will be discussed in the next section.

Dark Experiments Between O_3 and α-Pinene

As noted earlier, in the September 11, 1979, run, condensation nuclei were detected approximately 15 min after the addition of 0.06 ppm O_3 and 0.4 ppmC of α-pinene. The trace of the condensation nuclei counter (CNC) is shown in Figure 4. Note that the CNC trace returned to zero when the monitor was momentarily switched to the chamber half that contained neither α-pinene or O_3. A maximum number of 3000 nuclei per cc was recorded. Nucleation may have begun sooner than was observed, because given the length of the sampling manifold ($\sim$40 ft), a high potential exists for some manifold diffusion losses of the smallest particles in the CN range prior to entering the condensation nuclei counter. If one were to assume, based on the more detailed particle information given in the next experiment, that the average detected particle was nonetheless about 0.1 μm, then approximately 1% of the reacted carbon could be accounted for as CN aerosol. Doubling the particle diameter to 0.2 μm would give approximately a 10% carbon accountability.

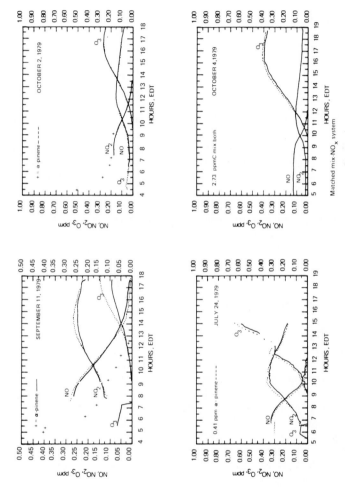

Figure 3. Effect of first reacting α-pinene and O₃ in matched carbon-NO$_x$ systems.

After 1 h of reacting 0.068 ppm of O_3 with 0.41 ppmC α-pinene, only 0.21 ppmC (0.021 ppmV) of α-pinene remained in the gas phase. Thus, 0.019 ppmV converted to gas phase products and aerosols during this time and 0.014 ppm O_3 disappeared. Of this amount, an O_3 decay rate calculation that was previously determined for low concentrations and dark phase, could account for a wall loss of only 0.003 ppmV/h. Hence, during the first hour, 0.02 ppmV of α-pinene and 0.01 ppm of O_3 had reacted. This amount falls within the lower end of the reaction stoichiometry previously cited by Ripperton et al. (1972), and is very similar to the value reported by Hull at this Symposium. A rate constant for the reaction between O_3 and α-pinene of 3.6×10^{-4} 1 mol^{-1} sec^{-1} was calculated from the net O_3 disappearance data. This number is slightly lower, though not very different, from other values which appear in the literature.

On the night of December 19, 1979, α-pinene and O_3 were reacted in the large UNC aerosol chamber (Fox et al. 1975) so that the aerosol equipment located in the laboratory beneath the chamber could be utilized to study the changing aerosol size distribution. Aerosol data were collected with an Environment/One CNC, a Thermo Systems, Inc. 3030 electrical aerosol analyzer (EAA), and a Royco 220 optical particle counter (OPC). A brief description of these instruments will follow to aid discussion of the experiment.

Condensation Nuclei Counter--

The CNC is used at UNC as an aerosol trend indicator. Absolute data collection is not possible because the physical measurement principles involved render primary calibration of the instrument very difficult. In addition, the range of particle sizes to which the instrument responds is not solidly established and probably varies between instruments. While the manufacturer claims a minimum detectable diameter of 0.002 μm, Liu and Kim (1977) have proposed a much larger diameter. Also, since no upper size limit of detection is supplied, this laboratory generally assumes 0.5 μm as an arbitrary upper limit.

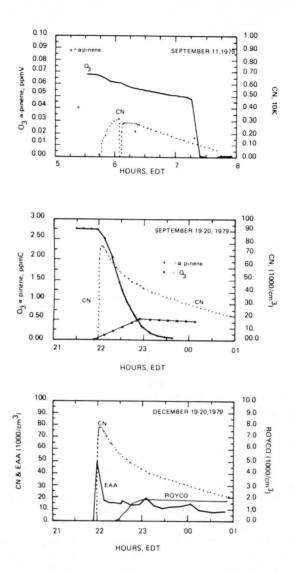

Figure 4. Dark reaction between O_3 and α-pinene and resultant total aerosol number traces.

Optical Particle Counter--

The OPC counts and determines the size of individual particles by a light scattering technique. A decrease in particle counts at the extreme lower detection limit is usually due to reduction of scattering intensity for particles in this region, rather than an actual decrease in particle count. The instrument used in this laboratory has an effective detection range of approximately 0.4 to 5.0 μm. Particle sizes are calibrated by generating monodisperse latex bead aerosols of known diameter. Latex beads of 0.4, 0.6, 0.88 and, 1.2 μm diameters were generated on December 4, 1979. Since the OPC was operated over a long period without extensive cleaning, a count calibration was also performed after the experiment. The possible associated count error could be very high.

Electrical Aerosol Analyzer--

The EAA operates on the principle of "diffusion charging - mobility analysis" explained by Whitby and Clark (1966), whereby instrument responses for particles between 0.003 and 1.0 μm are separated into ten channels. This allows data reduction to produce three-dimensional aerosol growth plots similar to those from the OPC, but lacking in resolution of particle size.

An inherent problem with the EAA develops from the fact that it actually measures the charge on particles, which varies with particle size. Data in the two lower channels are difficult to recover because only a small percent of particles carry a charge, and thus, the signal-to-noise ratio is small. Also, because larger particles carry varying amounts of charge, monodisperse aerosols can be distributed over many channels (Liu and Pui 1975). This laboratory has also often found a discrepancy between the EAA and OPC results in their respective crossover region (the last EAA channel). This discrepancy can occur if a large particle count appears in the middle channels, since hardware associated with analog-to-digital data conversion must be altered to relay large numbers to a data storage device correctly. This alteration lowers the count sensitivity by an order of magnitude, making channels with few counts (especially the high-size channels) prone to larger error. Errors are usually most evident on volume plots. Therefore, the EAA is an excellent instrument for describing the average aerosol size distribution at a given time in this size range (0.003 to 1.0 μm), but by virtue of its measurement technique, the calculated values from specific channels cannot be viewed as absolute (especially near either limit of detection).

Instrument Results--

After sunset on December 19, 2.75 ppmC equivalent liquid α-pinene was gently vaporized into the chamber at 2100 h; 0.5 ppm of O_3 was then ramp injected from 2155 to 2252 h. The concentrations of α-pinene and O_3, along with the total particle counts from each aerosol instrument, are shown in Figure 4. Within 5 min after the addition of O_3, aerosol was observed by the CNC and EAA meters. The slight difference in response between both of these instruments is due to differences in acquisition periods.

Inspection of the EAA number distribution vs. time plot (Figure 5) shows a large burst of particles with diameters near 0.01 μm immediately after the start of O_3 injection. Growth into the larger size ranges of the EAA (0.05 to 0.4 μm) took place over the next 30 min, with the greatest numbers appearing in the 0.07-0.13 μm range. The surface and volume distributions vs. time that are associated with the EAA detectable particles are shown in Figure 5.

Coincidental with the later stage of EAA growth, OPC particles began to accumulate (Figure 4) approximately 30 min after the start of the O_3 injection. Some of these particles have diameters within the EAA detection limits for the last channel (0.56 to 1.0 μm), even though no EAA counts were recorded there. This result is due to the previously stated sensitivity loss in the last channels due to high counts in the middle channels. Unfortunately, since the largest EAA particles are measured in these channels, a great deal of volume is not accounted for even though only a small percentage of the overall particle count is lost.

Because the EAA responded only to particles below 0.56 μm, the EAA and OPC have detected relatively exclusive sets of particles and the data sets are effectively isolated from each other. The volumes calculated from these two sets of data can be added to give an estimate of the total aerosol volume in both regions. The actual volume is expected to be slightly greater than the calculated volume, since the presence of any particles in the next to last EAA channel cannot be established. Thus, even though most of the particles occur in the middle of the EAA region, due to the larger size of the OPC particles, most of the total volume will reside in the optical region. An estimate of the total aerosol volume based on different calibration procedures ranged from ~500-2000 $μm^3/cm^3$. Of this amount, 97-99% were OPC-detected particles.

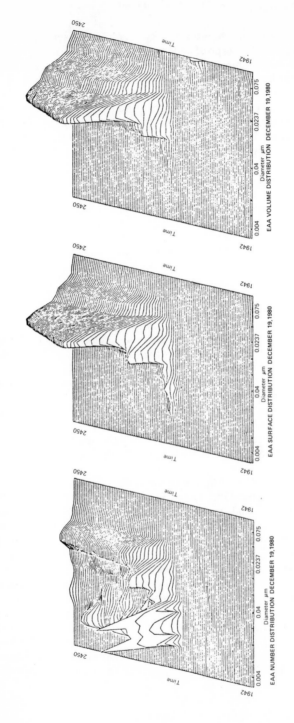

Figure 5. EAA three-dimensional number, surface and volume plots from the reaction between 2.75 ppmC α-pinene and 0.5 ppm O_3, December 19, 1979.

Assuming that the liquid aerosol product has a density of 1 g/cc, a crude conversion to aerosol mass loading can be made. This conversion permits comparison between the percent of carbon as α-pinene reacted, with the percent detected as aerosol. Our estimates suggest that the aerosol yield for this experiment was between 25% and 90%. Again, the high uncertainty of these numbers must be stressed. Hull (1980) suggested that some of the α-pinene O_3 reaction products have vapor pressures of 1 ppm. Since the smog chamber experiments were begun with 0.3 ppmV of α-pinene, if 10-40% of the products existed in the gas phase, then an upper limit on the total concentration of these compounds would be 0.1 ppmV. This concentration would permit many of these products to exist in the gas phase, as long as critical aerosol surfaces did not form and aid in the condensation of the compounds.

SUMMARY AND CONCLUSIONS

If α-pinene and O_3 reaction products are assumed to be photochemically inert, consideration of natural terpene emissions by O_3 control strategies is not necessary. Ambient levels of reacted terpenes are very low and the products of O_3 and terpenes would not significantly participate in the critical reactions of O_3 production.

On the other hand, if these products are as reactive as α-pinene or an urban mix, then continued refinement of natural emissions inventories and product measurements would be needed to determine the extent to use natural emissions in full-scale urban or regional photochemical models.

The substitution of 20% α-pinene carbon into a complex urban-like HC-NO_x system does not lead to discernible changes in NO_x or O_3 concentration rates and times. In the case of 20% α-pinene carbon addition, increases in O_3 concentrations and rates were similar to the enhancements that would be expected from the same carbon increase in the HC mix concentration.

From the experiments performed in this study, it is not clear how the products of terpene-O_3 reactions should be treated. Data have been presented suggesting that prior formation of these products may tend to reduce O_3 yields from urban-NO_x smog systems, when compared to the O_3 yields from unreacted terpenes within similar urban smog-type mixtures. Other experiments do not confirm these results but do indicate that the reacted products do not impart any "unusual" reactivity to the smog system.

Aerosol measurements were made during the dark reaction of O_3 and α-pinene and indicate that the majority of particles resulting from this process reside in the 0.1 μm diameter range. Most of this mass, however, was contributed by particles >0.6 μm diameter. Carbon accountability of reacted α-pinene in the product condensed phase was estimated to be 25-90%. As noted before, a very high uncertainty exists with this estimate.

REFERENCES

Fox, D. L., J. E. Sickles, M. R. Kuhlman, P. C. Reist, and W. E. Wilson. 1975. Design and operating parameters for a large ambient aerosol chamber. J. Air Pollution Control Assoc. 25:1049-1053.

Gay, W. B., Jr. and R. R. Arnts. 1977. The chemistry of naturally emitted hydrocarbons. In: Proceedings International Conference on Photochemical Oxidant Pollution and Its Control. EPA-600/3-001a. U.S. Environmental Protection Agency, Research Triangle Park, North Carolina.

Hull, L. 1981. Terpene ozonolysis products. In: Proceedings of the EPA Symposium on Atmospheric Biogenic Hydrocarbons: Emission Rates, Concentrations, and Fates. Ann Arbor Science Publishers, Ann Arbor, Michigan.

Jeffries, H. E., D. Fox, and R. Kamens. 1975. Outdoor Smog Chamber Studies: Effects of Hydrocarbon Reductions on Nitrogen Dioxide. EPA-65/3-75-011. U.S. Environmental Protection Agency, Research Triangle Park, North Carolina.

Jeffries, H. E., R. M. Kamens, and K. Sexton. In preparation. The Effectiveness of Reducing the Reactivity of an Urban Mix by Hydrocarbon Substitution.

Kamens, R. M. and H. E. Jeffries. 1979. Aspects of Natural Hydrocarbon Reactivity and Its Influence on Urban Smog. Environmental Sciences Research Lab Seminar Series, Research Triangle Park, North Carolina, Nov. 19.

Lonneman, W. A., R. L. Seila and, J. J. Bufalini. 1978. Ambient hydrocarbon concentrations in Florida. Environ. Sci. Technol. 12:459-463.

Lonneman, W. A. and J. J. Bufalini. 1979. Letter to the Editor. Environ. Sci. Technol. 13:236.

Liu, B. Y. H. and D. Y. H. Pui. 1975. On the performance of the electrical aerosol analyzer. J. Aerosol Sci. 6:249-264.

Liu, B. Y. H. and D. Y. H. Kim. 1977. On the counting efficiency of condensation nuclei counters. Atmos. Environ. 11:1097-1100.

Miller, P. R., J. M. Pitts, Jr., and A. H. Winer. 1979. Factors in summer ozone production in the San Francisco air basin. Science. 203:81-82.

Ripperton, L. A., H. E. Jeffries, and O. White. 1972. Advanced Chemistry. Ser. No. 113, p. 19.

Sandberg, J. S., M. J. Basso, and B. A. Okin. 1978. Winter rain and summer ozone: A predictive relationship. Science. 200:1051.

Westberg, H. and M. Holdren. 1976. Quarterly Report. Prepared for U.S. Environmental Protection Agency under EPA Research Grant No. 8000670-03. Research Triangle Park, North Carolina.

Whitby, K. T. and W. E. Clark. 1966. Tellus. 8:573-579.

Zimmerman, P. R. 1977. Procedures for conducting hydrocarbon emission inventories of biogenic sources and some results of recent investigation. Paper presented at U.S. Environmental Protection Agency Workshop on Emissions Inventory Factors, Raleigh, North Carolina.

Zimmerman, P. R. 1978. Response to Questions Raised by EPA-GKPB Review of NEDS Draft Final Report and Preliminary Florida Emissions Inventory. Special Report. Prepared by Washington State University, Pullman, Washington, for U.S. Environmental Protection Agency under Contract No. DU-77-C063.

DISCUSSION OF PRESENTATION

WALKER: You made that mixing at night. Did you see aerosol when shining a light into it?

KAMENS: No. There's not a long enough path length on the chamber. But that is an interesting question. It seems to me Don Fox ran some high concentration xylene-O_3 runs in his chamber. You actually could see a mist in the chamber. Is that correct?

VOICE FROM AUDIENCE: You had about 40 ppm?

KAMENS: Yes. Concentrations were very high.

VOICE FROM AUDIENCE: You can't see a decent path in the chamber.

HULL: Did you say, when you integrated all the particles, that you accounted for about 25% of the material as particle?

KAMENS: That's correct. Now, there are two assumptions that are made. First of all, we did not account for oxidation products. The molecular weight of α-pinene is 136; resulting products are going up to a molecular weight of, say about 170 or 182. So that is about a 11% or 12% increase? Well, it's a bigger increase than that. My feeling is that the density of α-pinene is about 0.86; the density of these things is greater. They approach 1. So these are sort of offsetting penalties.

So, in answer to your question, yes. According to this kind of calculation, we are looking at 25 or 30%. Or maybe even more. If we miss the calibration, based upon the lack of agreement between the Royco data and the electrical aerosol data, then we may be able to see more in the aerosol phase.

HULL: That is interesting because 25% is the fraction of material that I had. That was cis-pinonic acid which was the higher, least volatile, of all the materials.

OLLISON: Was there any reason that it should form particles in one case, and not in another? Temperature or something like that?

KAMENS: I looked very carefully at the conditions. The dew points on both of those data are similar; it's approximately 55°. It was cooler in the morning, and I might mention that for the run we made in which we were measuring aerosols in the Whitby as well as in the optical range, it was a cool evening. I think it was 45°F outside.

The problem is I could not see any gross differences. I couldn't see any fine differences between those two experiments which would make me want to throw one or the other of them out and say "Well, I'm suspicious." I am not suspicious, actually. I looked at the data very carefully and liked them all.

ALTSHULLER: Are you saying that the data you are putting together from various optical counters have a bimodal distribution?

KAMENS: If I believe the Whitby data, and if I believe the Royco data, it appears as though there is particle growth in two regions. We did some calculations; it doesn't take a lot because there are some larger particles that are forming. These would be below the lower detectable limit of the largest channel on the Whitby, or even the next largest; the 0.4, the 0.7 micron range that could creep on through, into the Royco range.

The rate needed in order to see the buildup was between one and ten particles per second per cc, moving through that range.

ALTSHULLER: Is there a mechanism of growth that would explain particles on that range? My recollection is, from what has been suggested in the literature, that it would be more likely that the particles would be in the tenths or two-tenths of a micron range, rather than a higher range.

JEFFRIES: Let me make a comment on that. Depending upon the nature of the condensing species of aerosol, the particles form in the small range, or seems to be the EAA range, or the Royco range. For cyclohexene, for example, there is no response with the CN counter, essentially no response on the EAA instrument; it's all in the optical instrument.

When I say "no response," I mean that about a couple of hundred particles per cc could be formed in moving through the system, and you would not be able to detect them with instrumentation. 1-hexene forms a smaller diameter aerosol exclusively, compared to cyclohexene, which forms essentially an aerosol only in the optical range.

So, depending on what the exact nature of the species is that is doing the condensing on the aerosol--to form the aerosol--you get it down here or up here, or whatever. One interpretation to be placed on this is that there's one set of species resulting in an aerosol in the EAA range, and as you get some coagulation into the larger size ranges, you suddenly reach a sufficiently low--high surface area, or low curvature--so that you condense out another species. And you can see that growth occurs only at the upper range. You can not see it in the lower range at all.

Thus, I think there are some mechanisms to account for the fact that you have aerosols in each mode, or one mode only, or in the other mode only.

KAMENS: An overview I did not show is the actual number counts versus time for each of the particle counts. The CN gave the highest count, but that doesn't mean that its count was correct; it's in the same order of magnitude as the Whitby.

But you can see that the Royco, which has the highest volume, has the lowest number of counts suggesting that something did happen to these smaller particles, that there was movement (either coagulation or condensation on top of these particles) to move them into the larger ranges.

OLLISON: Do you know of any evidence for isoprene particulate formation?

KAMENS: No. We haven't done any experiments with isoprene. We should.

OLLISON: Do you know of any literature?

KAMENS: I'm not familiar with that. But perhaps William Wilson could better answer that question.

W. E. WILSON: I don't know that anyone has ever looked at isoprene for particle formation. On the basis of work with other cyclic olefins and monoolefins, you would not expect particle formation of any significance.

ALTSHULLER: I think, again, if you worked at the higher concentrations with di-olefins, O_3, you could see aerosol.

JEFFRIES: As octadienates, to form octadienates on chains.

ALTSHULLER: No, I think we did some work several years ago with butadiene, again it may have been 40 ppm.

KELLY: Do you interpret your O_3 results to mean, in this mix system, that α-pinene looks a little more reactive than--in the pure system--it is shown to be by Arnts.

KAMENS: No, not at all. I think it's slightly more reactive, if you consider O_3 yields, than our atmospheric mix. If you compare propylene to our atmospheric mix--Arnts compared it to α-pinene--it's much, much more reactive in terms of conversion of NO into and with O_3 formation--

JEFFRIES: At 20% replacement level, α-pinene is like a mix; at 20% replacement level in the mix, it stands in, exactly the same as the mix.

SUMMARY COMMENTS

Joseph J. Bufalini

It appears that a great deal more work is needed, especially in obtaining good emission rates and biomass factors. However, I am not convinced from the presentations made in the last two days that natural hydrocarbons are necessarily important in the generation of significant levels of ozone. I base this opinion largely on ambient air measurements, chemical reactivity studies, and modeling efforts.

I would like to see more work done on product identification, especially on the aldehydes, ketones, and alcohols produced. Unfortunately, no good techniques are available for such measurements. Environmental Research and Technology, Inc. (ERT) has recently developed a new technique for aldehydes, but this technique is still untested.

Besides product identification, we plan to look into several areas within the next year. First, some carbon-14 work on aerosols is needed. This study should establish the biogenic contribution to the total carbon particulate in the atmosphere. Of course, size distribution of the particulates is necessary since pollen would give erroneously high carbon-14 values. Also, as Dr. Hull has suggested, some of the organic particulate may revolatilize. The work must then be such that revolatilization is minimized.

More work is needed in photochemical modeling. Different emission and meteorological scenarios could be established to ascertain the importance of biogenic emissions. The problem with this is that a good chemical model cannot be written. As Mr. Arnts pointed out with his modeling, propylene was selected as the surrogate terpene because he could not write a chemical mechanism for either isoprene or alpha-pinene. Even the propylene mechanism is

211

not completely understood, since the amount of ozone predicted from a propylene-NO_x system is slightly greater than the observed values. Unless we identify a large fraction of the products arising from the photooxidation of terpenes, it will be almost impossible to write a decent mechanism.

SUMMARY DISCUSSION

ALTSHULLER: If we are to construct a mechanism - isoprene shows a larger products identification. Why not write a mechanism for a smaller molecule like isoprene?

BUFALINI: I'm not certain we could write a mechanism even for isoprene.

ALTSHULLER: Since the isoprene molecule is small and little aerosol is produced, isoprene would be a better surrogate for the natural hydrocarbons.

ARNTS: Only 31% to 44% of the isoprene is accounted for.

ALTSHULLER: I have difficulty in understanding in this day and age why it's so difficult to write a mechanism for a simple molecule like isoprene.

BUFALINI: You are right, but I hasten to point out that we still have problems with toluene. O'Brian from Portland State and Hendry from SRI have been looking at this compound for several years; probably 5 man years have already been spent on it and we still cannot write a good mechanism.

ALTSHULLER: We can obviously debate endlessly on how much time it will take to understand the mechanism for any particular molecule.

BUFALINI: I thought it was interesting to note that Dr. Knoerr found haze in the wintertime. Was this also true in Russia, Dr. Wilson? [Dr. Wilson and Mr. Stevens studied the formation of haze in Russia].

W. E. WILSON: I don't know, but I will inquire.

BUFALINI: It's not impossible since there is always some stratospheric ozone, around 20 to 30 ppb, and this could react with the terpenes emitted from conifers. However, I am not certain that there is sufficient terpene to react with

213

the ozone to produce the reduced visibility (produce aerosols).

W. E. WILSON: We must remember that in these haze situations, we are frequently looking at vistas of 30 to 60 miles away. So, we aren't talking about large concentrations of aerosols in these situations.

OLLISON: In estimating natural emissions for urban areas, factors of 50% to 70% show up as natural compared to anthropogenic; for nonurban areas, the percent of natural is a bit greater. Conventional wisdom has been that these natural hydrocarbons are very reactive and they go to aerosols which removes them from the mechanism.

There seems to be a little more doubt about that after this symposium. We have no feel as to where this tremendous estimated quantity of hydrocarbon is going.

BUFALINI: If we don't assume that the natural hydrocarbons are producing aerosols, then where is the material going? It does not appear to be in the gas phase.

OLLISON: The second-to-the-last speaker [Dr. Hull] sees a lot of oxygenated products; he is having to go to close to 300°C on his column. It's not that they [the oxygenates] are not there, but that we haven't been sufficiently careful to analyze for them.

ALTSHULLER: I think you're losing the point. The terpenes react faster than something like isobutene--perhaps by a factor of two with ozone. However, you can see things like isobutene and other reactive species in cities hour after hour during the day. Why can't we see these natural hydrocarbons if the emissions are as high as 50% to 70% of the anthropogenic? It seems inconsistent.

OLLISON: If the natural hydrocarbons are comparable to the anthropogenic, and there are four for each anthropogenic--the problem then is relying on just control of the anthropogenic and there are, say, four of those for each anthropogenic one, the problem is relying on just the control of the anthropogenic. I guess I am missing the point.

DIMITRIADES: Another way you can define this is to use emissions data in the simulation models. These should be used for both anthropogenic and natural emissions. The emission rate measurements are high--or so they appear--the question is what emission rates does one use? Did you use those rate data or what? What emissions data are you using in the model?

BUFALINI: Even if we agree with the Zimmerman calculation that 68% of the total hydrocarbon in Tampa-St. Petersburg is from vegetation, the concentration of natural hydrocarbon on a part per billion basis is still low. Pat Zimmerman, as I recall, had calculated low ambient levels of natural hydrocarbons using his emission rates. I don't remember the exact figure.

VOICE FROM AUDIENCE: I think the calculation showed 20 ppb to 30 ppb.

BUFALINI: I thought Zimmerman's calculation showed natural hydrocarbons were lower than that. However, even so, this isn't very much natural hydrocarbon. However, if we fly in an aircraft over the Tampa area, say 300, 500, or 1000 ft, the TNMHC is between 70 ppb and 140 ppb. If we look at the gas chromatogram, the hydrocarbons are, for the most part, indicative of dilute auto exhaust.

DIMITRIADES: What is your point? There is no evidence of significant amounts of natural organics in the air?

OLLISON: No one knows where the natural hydrocarbons are going.

DIMITRIADES: You see, the model should use emission rates, not ambient concentrations.

BUFALINI: But both of you are ignoring what Dr. Altshuller said.

ALTSHULLER: Yes, are they disappearing like magic or science? If they are disappearing by science, then what are the terpenes being consumed by, and why aren't other hydrocarbons being consumed? The reactive species must be ozone and hydroxyl radical.

DIMITRIADES: What you're saying is that the emissions data are wrong?

ALTSHULLER: Yes, as far as I can tell.

DIMITRIADES: What then? What do we do with the models? We have to use a number.

ALTSHULLER: Use the ambient air data and work backwards with the model.

BUFALINI: It seems like a viscious circle. Usually, one uses the ambient data to validate the model. In this case, we must have a valid model and then use ambient data to

verify the emissions. We were hoping to do this in the
Tampa—St. Petersburg area since we have the emissions data.

OLLISON: The dichotomy is that modelers like to use
emissions data, whereas scientists like to use the
atmospheric concentrations in the model.

BUFALINI: Dr. Demerjian had some problems when he ran the
model in St. Louis. The ambient air data did not agree with
the emissions data.

JEFFRIES: Dr. Demerjian just adjusted his aromatic content
in the St. Louis data base from something like 10% to 18%.
Just on the basis of the way the emission estimates were
done, there could easily be a 10% error on the amount of
aromatics. The correction was based on Lonneman's data for
the area, which showed that there was about 70% aromatics in
the ambient air mix.

DIMITRIADES: If you're concerned about reconciling the
emissions data with the ambient concentrations, you should
look at the ratio of emission and the ratio of the ambient
air. This is where the anthropogenic emission question comes
up. But, if you're interested just in the emission rate data
for the natural organics, it has nothing to do with the
anthropogenics. Only if you look at the natural organics
relative to the anthropogenic, and only then would you be
interested in the anthropogenic emission rate.

BUFALINI: But the ambient air data show that the Zimmerman
emissions factor must be reduced by a factor of 10 to 15. We
don't see the concentrations predicted by the simple
diffusion model by using the Zimmerman emissions.

DIMITRIADES: The suggestion that we go backwards from the
ambient concentrations of the natural organics and then try
to estimate emission rates appears to be the only
constructive suggestion to resolve the issue.

However, there is another question that needs to be
addressed. If we put natural organics in the model, do we
use the same chemical model? Don't we need to revise the
model to take care of the aerosol products, etc.? Perhaps
for the first solar day this may not be crucial, but for
multiday reactions, this might make a difference.

ROMANOVSKY: Given the basic uncertainty of the air quality
simulation models, where most modelers hope for ∓ 30%
uncertainty, would you hope and expect to see any impact of
2% or 4% of the hydrocarbon as natural hydrocarbon?

DIMITRIADES: With the natural hydrocarbon only 2% to 4%, no.

ROMANOVSKY: Mr. Seila's presentation on the Jones State Forest showed only 2% to 4% of the hydrocarbons in the canopy as having a natural origin. Then, this low percentage would have little or no effect when the hoped-for accuracy is only $\mp$ 30%. Therefore, I don't see the relevance of trying to fit in these natural hydrocarbon contributions.

JEFFRIES: I think there are two issues. One is that most modelers do not want to model 2-day or multiday phenomena. This is because there is plenty to worry about for the one day. There are many issues in the nitrogen chemistry that have to be resolved before we can talk about what organics are going to do on the second day. If you look at smog-chamber results, up to 20% (natural hydrocarbons) there isn't much difference.

There is also the interface between the ozone and nitrogen oxides chemistry. It's one thing to predict ozone and do a good job of it, but it's something else to have other modeled components agree with the experimental measurements. You can force the model to give you ozone with a reasonable fit, whereas the PAN, the nitric acid, and the amount of organic nitrates are totally wrong. We are going to have problems when we run complex hydrocarbon mixtures. What happens to the organic nitrates? The longer the chain, the more stable the nitrates. Instead of oxidizing NO to NO_2 by the RO_2 radicals, the reaction goes to organic nitrates. As the chains get longer, less nitric acid is produced and organic nitrates increase.

BUFALINI: Dr. Rasmussen, you stated earlier that we are preoccupied with isoprene and alpha- and beta-pinenes, and we should be looking at the oxygenates that are emitted such as camphor. Do you think we should spend more effort on oxygenates liberated by vegetation and why?

RASMUSSEN: The first point I made was that we cannot leave Zimmerman's estimates where they are, since I would measure isoprene, alpha- and beta-pinene, myrcene, and limonene, while I would not put a name on the rest of the material.

Two, it's at a point where you need much better measurements of the ambient atmosphere in rural sites where, instead of trying to make ten measurements a day, you collect one or two very well-integrated samples. Process that air so thay you come up with a concentrate that you can make multiple analyses on and begin to really identify with GC/MS

217

what it is specifically you're looking at. That in the past
has been shelved as noise.

I certainly do not accept Bob Steven's chromatogram from
Russia. It doesn't bear any resemblance to the kind of
chromatogram that you get in other remote parts of the world
for the background level of benzene and other trace gases. I
think that's the approach. We have to back-estimate from
what we observe in the atmosphere on how well this fits with
these emissions.

We know too well that anytime you enclose the foliage,
you run the risk of disturbing that very delicate environment
with the stomatal resistance and thermal feedback and
incipient wilt. This most likely potentiates a lot of
hydrocarbon emission that you measure in the tissues.

BUFALINI: How would you respond to Mr. Romanovsky's point of
2% to 4% not being important?

RASMUSSEN: I'm not convinced that there's only 2% to 4%
natural hydrocarbon in the canopy of the forest. I'm
convinced that there's much more than that in the canopy.

WALKER: The 2% to 4% was a January measurement, which could
be low compared to summer numbers.

RASMUSSEN: We measure much higher things than that in the
canopy. It's too bad Zimmerman has already left, because I
would like to contrast the measurements I have and those he
has with what is found in the tropics.

ALTSHULLER: Well, during the year, what about the North and
South Poles? Let's at least stay in the temperate zone.

RASMUSSEN: I think you have to have a bit of comparison in
order to appreciate what the temperate zone really means.

ALTSHULLER: That's a nice generality, but in terms of
addressing air pollution problems, the question is what do we
really have to do to get the problem better pinned down.

BUFALINI: We're really not in disagreement since we observed
as much as 60% of the TNMHC as natural at Chickatawbut Hill
southwest of Boston. This was a deciduous type of forest.
We also found ~50% of the TNMHC from natural sources at the
IBP site in North Carolina. However, in both cases the total
amount of hydrocarbon is low--usually below 100 ppb.

ALTSHULLER: This is within the canopy. Each time the natural hydrocarbon made up a substantial part of the total, we find that the total is quite low. Is that correct?

BUFALINI: Yes, the total was low. You agree with that Rei [Dr. Rasmussen]?

RASMUSSEN: Yes, I'd agree with that.

ALTSHULLER: I think that this is an important clarification.

RASMUSSEN: I also think more work should be done on the emission measurements from the foliages, similar to the type of work David Tingey was doing with less enclosure around the foliage and less thermal sealing in a plastic bag or in a glass bottle. Try to get emissions from a system like an aircraft wind tunnel where the system is sitting in an open tube and you look at the quality of air going in and quality of air going out and make certain that the dynamics over the leaves are not different from what you have in the real atmosphere. Under these least disturbing conditions, one can pick up the emissions from the foliage.

You're going to be surprised because we did this initially. Zimmerman, Mike Holdren, and I on a Coordinating Research Council (CRC) project hardly measured anything. We had to slow the air flow down or we had to process a much larger air sample.

I think the emissions, the natural organic emissions from plants are going to be a lot lower than what you get from these bags. My personal feeling and experience is that the emissions are going to be much, much lower.

The point I made in the first slide was that when you compare the emissions of isoprene from the Zimmerman dynamic bag system versus just a plastic bag versus the enclosures that Dave Tingey was using, as well as Al Jones' work just punching out leaf discs, and even the measurements of Sanadze's of a sweating plant over in Tbilisi—we all come up with the same order of magnitude: a pretty high emission rate of about 10 to 30 µg/g of leaf tissue/h. This is a very high rate. I think we've got to go back to more gentle study techniques and try to approach a more undisturbing measurement method. This may require much more laboratory work and less work in the field.

WALKER: Dr. Bufalini, it appears from Kamens' report that you must regard the natural hydrocarbons as about equivalent in average mix and reactivity as far as their oxidation in the real world when mixed with ordinary anthropogenic hydrocarbons.

KAMENS: Yes, but I would add that if you're only going to add 3% or 4% in carbon to any system, there would be almost no change.

WALKER: But it certainly is not in line with the statement that it's destroying ozone or lowering the--

JEFFRIES: No, that doesn't seem to do - it can do either.

WALKER: Well, I would agree that it could go either way under certain conditions. They do react but I was a little surprised at how slow your reaction was between pinene and ozone.

JEFFRIES: Yes, but look at the levels - they are going to be even lower, resulting in even slower reactions in the real world.

I would also like to comment on a measurement that is probably not in the literature. I picked it up when I was in Sydney, Australia. At night, drainage flow from the eucalyptus forest goes into Sydney. They are looking into the natural hydrocarbon impact on Sydney's oxidant problem. Using freeze-out techniques and detailed chromatography, they didn't find more than a tenth of a ppbC terpene material. This is consistent with what we have been saying here. The emissions are much lower than the emissions measurements we suggest.

It looks like the whole problem is ambient measurements versus leaf measurements. We should get them together in some way.

BUFALINI: Our group (GKPB) is planning to do more emission measurements work. We also plan to do some additional photochemical modeling and we must spend more time on mechanistic studies. I'm also very impressed with Les Hull's work, but this is at very high concentration. This bothers me because the levels are much lower in the atmosphere.

JEFFRIES: His work is also in the absence of nitrogen oxides.

BUFALINI: There is some evidence to suggest that the products are different when the concentrations are high. This was found to be true with H_2O_2 - it's not produced when formaldehyde is photodissociated at high concentrations.

I would like to thank everyone for coming.

(Conference was adjourned)

APPENDIX

DISCUSSION PARTICIPANTS

A. P. Altshuller, Director
Environmental Sciences Research Laboratory, MD-59
U.S. Environmental Protection Agency
Research Triangle Park, North Carolina   27711

Viney Aneja
Northrop Services, Inc.
Post Office Box 12313
Research Triangle Park, North Carolina   27709

Robert Arnts
Gas Kinetics and Photochemistry Branch
Environmental Sciences Research Laboratory, MD-84
U.S. Environmental Protection Agency
Research Triangle Park, North Carolina   27711

Elgene O. Box
Department of Geography
University of Georgia
Athens, Georgia   30602

Joseph J. Bufalini, Chief
Gas Kinetics and Photochemistry Branch
Environmental Sciences Research Laboratory, MD-84
U.S. Environmental Protection Agency
Research Triangle Park, North Carolina   27711

George Dawson, Atmospheric Scientist
University of Arizona
Tuscon, Arizona   85721

Robert Denyszyn, Manager
Research and Development
Scott Speciality Gases
Route 611
Plumsteadville, Pennsylvania   18949

Basil Dimitriades
Environmental Sciences Research Laboratory, MD-59
U.S. Environmental Protection Agency
Research Triangle Park, North Carolina 27711

Jack Durham
Environmental Sciences Research Laboratory, MD-57
U.S. Environmental Protection Agency
Research Triangle Park, North Carolina 27711

Martin Ferman
Environmental Science Department
General Motors Research Laboratories
Warren, Michigan 48090

Dennis Fitz
State-Wide Air Pollution Research Center
University of California - Riverside
Riverside, California 92502

Don Flyckt
Washington State University
Pullman, Washington 99163

T. E. Graedel
Bell Laboratories
Murray Hill, New Jersey 07974

Michael W. Holdren
Battelle Columbus
505 King Avenue
Columbus, Ohio 43201

Leslie Hull
Department of Chemistry
Union College
Schenectady, New York 12308

Harvey Jeffries
School of Public Health
University of North Carolina
Chapel Hill, North Carolina 27514

224

Murray Kaiserman
Northrop Services, Inc.
Post Office Box 12313
Research Triangle Park, North Carolina 27709

Rich Kamens
School of Public Health
University of North Carolina
Chapel Hill, North Carolina 27514

Nelson Kelly, Associate Senior Research Scientist
Environmental Science Department
General Motors Research Laboratory
Warren, Michigan 48090

Kenneth R. Knoerr
School of Forestry and Environmental Studies
Duke University
Durham, North Carolina 27706

David Lincoln
Department of Biological Sciences
Stanford University
Stanford, California 94305

William Lonneman
Environmental Sciences Research Laboratory, MD-84
U.S. Environmental Protection Agency
Research Triangle Park, North Carolina 27711

Kenneth H. Ludlum, Research Associate
Texaco, Inc.
Post Office Box 509
Beacon, New York 12508

Charles O. Mann
Office of Air Pollution and Quality Control, MD-14
U.S. Environmental Protection Agency
Research Triangle Park, North Carolina 27711

Will M. Ollison
American Petroleum Institute
2101 L Street, N.W.
Washington, DC 20037

Reinhold A. Rasmussen
Oregon Graduate Center
19600 N.W. Walker Road
Beaverton, Oregon 97005

J. C. Romanovsky
Environmental Sciences Research Laboratory
U.S. Environmental Protection Agency
Research Triangle Park, North Carolina 27711

Robert L. Seila
Environmental Sciences Research Laboratory, MD-84
U.S. Environmental Protection Agency
Research Triangle Park, North Carolina 27711

Robert Sievers
Department of Chemistry
University of Colorado
Boulder, Colorado 80309

Robert K. Stevens
Environmental Sciences Research Laboratory
U.S. Environmental Protection Agency
Research Triangle Park, North Carolina 27711

Boyd Strain
Department of Botany
Duke University
Durham, North Carolina 27706

David T. Tingey
Environmental Research Laboratory
U.S. Environmental Protection Agency
Corvallis, Oregon 97330

Harry Walker
Monsanto Company
Post Office Box 711
Alvin, Texas   77511

Carol G. Wells
U.S. Forestry Sciences Laboratory
Post Office Box 12254
Research Triangle Park, North Carolina   27709

Hal H. Westberg
Washington State University
Pullman, Washington   99163

William E. Wilson
Environmental Sciences Research Laboratory, MD-84
U.S. Environmental Protection Agency
Research Triangle Park, North Carolina   27711

Patrick Zimmerman
National Center for Atmospheric Research
Post Office Box 3000
Boulder, Colorado   80307

INDEX

Abastumani Mountains
  aerosols, 99, 100
  volatile organics, 98, 99
Acetylene, 22
Aerosol formation, 27, 92-98,
  139-153, 157-159, 198-203,
  207, 208
Ambient air analysis, 54-61
Ambient hydrocarbon concentrations, 66-69

Camphor, 23, 24
Chamber contamination, 73,
  74, 136

Ethane, 1-15

Glacial hydrocarbons, 77-80
Great Smoky Mountains
  aerosols, 92-98
  volatile organics, 90-92

Houston (Texas), 1-24
3-Hydroxynopinone, 176, 179

Isobutane, 1-15
Isoprene, 11
Isoprene emissions
  seasonal variation, 37

Jones State Forest, 1-13

3-Ketonopinone, 178

Myrtenol, 178

Natural gas, 6, 67
Nopinone, 178, 179
Norpinonic acid, 163, 165,
  174

Oxygenated hydrocarbons,
  38-43

PAH's
  in glaciers, 79
Photochemical modeling,
  126-130, 133
  photochemical reactiv-
    ity, 69-71
  effect of water vapor,
    141-153
α-pinene, 11
  aerosol formation, 139-
    153, 157-159, 198-
    203, 207, 208
  ozonolysis, 161, 167-
    170, 180, 181
  photochemical oxidant
    potential, 121-135,
    187-209
β-pinene
  aerosol formation, 139-
    153, 157-159
  ozonolysis, 175, 180,
    181
Pinocarveol, 178
Pinonaldehyde, 163-165,
  172, 173, 179
Pinonic acid, 163-165,
  172, 174, 179
Ponderosa pine, 35